T0361842

Biochar and Soil Biota

Biochar and Soil Biota

Editors

Natalia Ladygina
Diepenbeek
Belgium

Francois Rineau
Hasselt University
Hasselt
Belgium

CRC Press
Taylor & Francis Group
Boca Raton   London   New York

CRC Press is an imprint of the
Taylor & Francis Group, an **informa** business

A SCIENCE PUBLISHERS BOOK

CRC Press
Taylor & Francis Group
6000 Broken Sound Parkway NW, Suite 300
Boca Raton, FL 33487-2742

© 2013 Copyright reserved
CRC Press is an imprint of Taylor & Francis Group, an Informa business

Cover Acknowledgement: Reflection Electron Microscope (REM) of original Terra Preta charcoal (Brazil). Reproduced by kind courtesy of Katja Wiedner, Martin-Luther University, Halle-Wittenberg, Germany.

No claim to original U.S. Government works

Printed in the United States of America on acid-free paper

International Standard Book Number: 978-1-4665-7648-3 (Hardback)

Visit the Taylor & Francis Web site at
http://www.taylorandfrancis.com

CRC Press Web site at
http://www.crcpress.com

Science Publishers Web site at
http://www.scipub.net

Preface

Biochar, a biomass that is burned at very high temperature in the absence of oxygen, has recently become an interesting subject of study. Biochar is highly stable and does not degrade; it possesses physical properties that assist in retention of nutrients in the soil. The use of biochar will undoubtedly have a significant impact not only on soil nutrients but also on soil organism communities and their functions.

This book focuses on how the ecology and biology of soil organisms is affected by the addition of biochar to soils. It takes into account direct and indirect effects of biochar addition to soils, on the soil carbon cycle, impact on plant resistance to foliar and soilborne disease, interactions with pathogenic, mycorrhizal and saprophytic fungi. The stability of biochar in soil environment is also discussed. Special focus has been put on application of biochar to remediate polluted soils, taking into account possible toxic effects of biochar on soil fauna. One of the chapters provides extended information on the role of biochar in soil using isotopic tracing techniques. An effort has been put to summarize the methods of traditional production of biochar, their material characteristic and how different biochars may change when applied to soils. Additional chapter is devoted to a comparison of methods to apply biochar into temperate soils.

The chapters have been written by experienced and internationally recognized scientists in the field.

This book will be useful to students and researchers in agronomy, biology, ecology, and environmental managers from both academic as well as industrial organizations.

Natalia Ladygina and Francois Rineau

Contents

The Stability of Biochar in the Environment

Andrew R. Zimmerman[1,*] and Bin Gao[2]

Introduction
 1 Biochar Recalcitrance and Lability
 2 Biochar Loss—Rates and Models
   2.1 Short-Term Laboratory Incubations
   2.2 Biochar Mineralization Models
   2.3 Field Studies
 3 Mechanisms of Black Carbon and Biochar Loss
   3.1 Biotic Black Carbon Degradation
   3.2 Abiotic Oxidative Black Carbon Degradation
   3.3 Non-Oxidative Abiotic Black Carbon Loss
   3.4 Black Carbon Loss by Leaching
   3.5 Black Carbon Losses by Erosion/Translocation
   3.6 Black Carbon Losses by Later Fires
 4 Biochar-Soil C Interaction and Stability
 5 Mechanisms of Biochar Stabilization
Conclusions and Summary
References

[1]Department of Geological Sciences, University of Florida, 241 Williamson Hall, PO Box 112120, Gainesville, FL 32611-2120; E-mail: azimmer@ufl.edu
[2]Department of Agricultural and Biological Engineering, University of Florida, Gainesville, FL 32611; E-mail: bg55@ufl.edu
*Corresponding author

Introduction

Pyrogenic organic matter (OM) is a carbon-rich material that is present in a variety of forms ranging from lightly charred biomass to charcoal to soot (Masiello 2004). It is known as 'black carbon' (BC) by geochemists and has come to be referred to as 'biochar' when created by pyrolysis in oxygen-limited conditions. An understanding of the effects of biochar amendment on soil nutrient cycling, water balance, ecology, soil fertility and other associated beneficial properties is still emerging. However, it has long been clear that biochar, as a relatively refractory form of OM, can be used as a carbon (C) sequestration tool. That is, biomass C in its pyrogenic form is less susceptible to remineralization (i.e., conversion back to CO_2 and perhaps CH_4) in the environment than its non-pyrogenic form.

Understanding and quantifying the longevity of different types of biochar in the soil environment is important for a number of reasons. First, making up between 5–30% of soil and sediment organic C (Skjemstad et al. 2002, Song et al. 2002, Masiello and Druffel 2003), BC represents a large but poorly quantified portion of the surficial global carbon cycle. Atmospheric CO_2 concentrations and associated climate changes of the past, as well as the historical record of fire occurrence, may be linked to the stability of BC in the environment. And any changes in the incidence of fire that occur with global climate change may, through BC, represent either a positive or negative climate change feedback, depending upon BC-soil-fire dynamics. Second, one who amends a soil for the purpose of fertility enhancement with a specific biochar type surely would wish to know how long those benefits could be expected to last. Some of these benefits are likely related to total soil C content, while other benefits may be related to other characteristics of biochar which may change over time (Ding et al. 2010, Graber et al. 2010, Mukherjee 2011). Lastly and most relevant to the present discussion, the large scale adoption of biochar soil amendment could potentially offset a substantial portion of the C released by humans through the burning of fossil fuels and thus serve as an important climate change mitigation tool. A recent estimate based on available waste biomass quantities, available land not already dedicated to food production, and C conversion efficiencies, predicted that biochar could offset up to 12% of annual net CO_2-C equivalent emissions (Woolf et al. 2010). Additional benefits may be derived from improved water use efficiencies and, should C sequestration activities be assigned a monetary value such as by C offset trading, rural economies could benefit as well. For this to occur, however, an accurate and long-term accounting system of the effect of biochar amendments of specific types on soil C would need to be developed.

This chapter is intended to review current knowledge on the stability of soil-amended biochar C as well as its effect on non-biochar soil C. Aspects of this subject have been reviewed previously (Lehmann et al. 2009, Spokas 2010, Knicker 2011). This work will focus on more recent advancements that have been made in this area, on topics not covered by those reviews, and will present previously unpublished data on these topics generated in our laboratories. We have also assembled data summary tables to compare rates of biochar C losses that have been measured in the field and laboratory by various means. The different measurement approaches will be evaluated in addition to models used to simulate degradation data and predict future C losses.

Although the focus of this chapter is on carbon, it should be understood that, as biochar is made up of 50–90% C by weight, the longevity of biochar C can represent the longevity of biochar as a whole. Also, the term biochar is used here when referring to pyrogenic material produced by humans, the term black carbon (BC) for pyrogenic material produced in nature. The terms are used somewhat interchangeably, however, when discussing generic concepts of pyrogenic OM.

1 Biochar Recalcitrance and Lability

According to Sollins et al. (1996), 'stabilization' of an organic material refers to its resistance to loss via degradation, erosion or leaching. The latter two processes will be discussed later, while biochar degradation or 'mineralization', i.e., conversion to an inorganic carbon form such as CO_2 via abiotic or cellular respiration processes, is discussed here. The presumption of biochar's or BC's relative resistance to degradation, or 'recalcitrance', in soil rests on a number of observations. Simplest is the very presence of old charcoal in soil. Soil charcoal ^{14}C dates can be thousands of years in age, much older that the organic carbon found in the same soil horizon (e.g., Glaser et al. 2001, Pessenda et al. 2001, Schmidt et al. 2002, Pessenda et al. 2005). *Terra preta*, BC-rich soils produced by pre-Columbian Amazonans through some method of burning, are surrounded by tropical soils with very little organic C, thus, are further testament to the relative recalcitrance of pyrogenic C.

The chemical recalcitrance of BC that has been observed in the laboratory also implies longevity in the environment. First, biochar is relatively resistant to acid, base and oxidant extraction. Thus, removal of non-BC carbon by chemical extraction followed by C analysis has long been used to quantify soil and sediment BC (Kuhlbusch 1995, Gustafsson and Gschwend 1998), though more recent research has found significant C losses using harsher extractants (Kurth et al. 2006, Hammes et al. 2007). Second, the molecular structures making up BC are known to be relatively resistant to microbial enzymatic attack (further discussion below).

On the other hand, some older and many recent studies have shown that BC degradation does occur. These include laboratory incubation experiments, measurements indicating C loss in the field over time, and observations of a degree of BC solubility. Each will be discussed in greater detail below and the various processes for C loss from biochar-amended soils compared. Assuming today's BC production rate via natural biomass burning of 50–270 Tg C year[1] (Kuhlbusch and Crutzen 1995), it has been calculated that 25–125% of the total soil organic C would have been produced as BC in just 20,000 years (Masiello and Druffel 2003). It is clear, then, that there must be BC losses, otherwise soil carbon would now be primarily BC. Czimczik and Masiello (2007) have termed this apparent contradiction the 'paradox of refractory-labile BC'. A solution to this paradox is surely in the heterogeneous chemical composition of biochar, both within any particular biochar type and across the different forms of biochar produced by different methods and from different feedstocks.

Molecular analysis of biochars reveals that, with heat treatment, or 'charring', plant biomass undergoes chemical transformations toward progressively more refractory molecular structures. Elemental analysis, and solid-state ^{13}C nuclear magnetic resonance (NMR) and Fourier Transform Infrared (FT-IR) spectroscopy have shown that, starting at temperatures as low as 250°C, the cellulose, pectin and lignin that make up the majority of plant dry biomass, is altered, predominantly by dehydration, to phenol, furan, aromatic and some alkyl C structures with substantial O, H and S substitutions (Baldock and Smernik 2002, Knicker 2007, Knicker et al. 2008, Keiluweit et al. 2010). At higher temperatures of 400–500°C and beyond, progressive depolymerization, loss of functional groups, aromatization, dehydrogenation reactions and removal of substituents occurs, resulting in larger sheets of fused aromatic rings. These condensed structures would likely be very resistant to biotic or abiotic degradation, as would the graphitic 'turbostratic crystallites' that have been observed to form at temperatures >600°C (Kercher and Nagle 2003, Keiluweit et al. 2010). Thus, the idea of a 'combustion continuum', attributed to Goldberg (1985) and popularized by Masielo (2004), was introduced, representing the degree of condensation and range in chemical structures represented by BC and biochar, with increasing temperature from slightly charred biomass to soot, and by implication, the range in environmental recalcitrance.

While the above discussion implies that production of recalcitrant molecules via heating is the primary cause of biochar's recalcitrance, the resistance of organic materials to abiotic or biotic degradation depends on a great variety of factors. In fact, the very idea of 'molecular-recalcitrance', as reviewed by Kleber (2010), is one that is only a "semantic convenience and not a useful classification of material properties". The resistance of biochar to degradation will likely vary with many additional intrinsic properties such as macro- and micro-pore structure, solubility, surface affinity for other soil components and extrinsic or environmental factors, such as pH, oxidant concentration, moisture level and soil composition and structure. The idea of biochar stability in the environment will evolve only as the understanding of its molecular composition grows, but also as we understand the biogeochemical interactions of each of its components with the environment in greater detail.

2 Biochar Loss—Rates and Models

2.1 Short-Term Laboratory Incubations

There have been a large number of recent publications investigating the stability of biochar. The great majority of them employ *in vitro* laboratory incubations of biochar alone or in soil and measure carbon dioxide (CO_2) efflux into the headspace of incubation chambers. Incubations of biochar with soil measure net C efflux from both the biochar and the soil and include the effects of soil-biochar interaction. These will be discussed separately. Results of a number of studies of these types that have examined mineralization of biochar-alone are shown in Table 1. Most commonly, these experiments were performed in cleaned quartz sand brought to 65% water holding capacity (WHC), were inoculated with a soil microbial consortia and a nutrient solution, and were incubated at 30°C in the dark. Conditions departing from these are indicated in Table 1. In order to calculate the biochar C mean residence times (MRT) and half-lives listed in Table 1 from the cumulative amount of effluxed CO_2, an exponential (first-order) decay pattern was assumed such that:

$$\text{BC mineralized}_{\text{(time t–end of incubation)}} = \text{BC}_{\text{(initial)}} (1-e^{-kt}) \qquad \text{(Eq. 1)}$$

where k is the apparent first-order degradation rate constant with units of inverse time. The MRT can then be calculated as $1/k$ and C half-life as $\ln(2)/k$. This has been called a '1G degradation model' because it assumes that the material, biochar in this case, is composed of one component degrading at a single rate.

The amount of biochar observed to have been mineralized in microbial incubations ranges from 0.3% (for oak wood pyrolyzed at 800°C for 24 h incubated over 60 d, Hamer et al. 2004) to 22.2% (for corn pyrolyzed at 250°C and incubated over 1 y at 60°C, Nguyen and Lehmann 2009). Calculated MRT of biochar C ranges from 3.1 y (for wheat straw pyrolyzed at 525°C for a few seconds incubated over 65 d, Bruun et al. 2012) to 658 y (for oak wood pyrolyzed at 650°C for 72 h and incubated over more than 3 y, extension of Zimmerman 2010 study). The data in Table 1 illustrate a number of trends that have been consistently observed in these types of studies. First, and

Table 1. Summary of results of studies measuring biochar mineralization in laboratory incubation by CO_2 efflux.

Source	Biochar Type (feedstock & production conditions)	Incubation Method[1]	Incubation period (d)	Biochar C (weight %)	Biochar C% mineralized[2]	Biochar C MRT[3] (y)	Biochar C Half-life[3] (y)
Baldock and Smernik 2004	pine sapwood, 200–350°C, 72 h, w/O_2	25°C	120	47.9–67.3	<2%		
Hamer et al. 2004	maize, 350°C, 2 h in muffle furnace in closed steel containers	20°C	60	66.4	0.8	21.0	14.6
"	rye straw "	20°C	60	66.3	0.7	22.7	15.8
"	oak wood, 800°C, 20–24 h	20°C	60	78.5	0.3	63.1	43.8
Nguyen and Lehmann 2009	corn residue, 350°C, 60% WHC	C loss determined by TOC difference	365	67.5	21.2	4.2	2.9
"	corn residue, 600°C, 60% WHC		365	79.0	11.2	8.4	5.8
"	oak wood, 350°C, 60% WHC		365	75.9	8.1	11.8	8.2
"	oak wood, 600°C, 60% WHC		365	88.4	8.9	10.7	7.4
"	corn residue, 350°C, 100% WHC	"	365	67.5	10.9	8.7	6.0
"	corn residue, 600°C, 100% WHC	"	365	79.0	9.4	10.1	7.0
"	oak wood, 350°C, 100% WHC	"	365	75.9	6.2	15.6	10.8
"	oak wood, 600°C, 100% WHC	"	365	88.4	8.6	11.1	7.7
Zimmerman 2010	Oak, 250°C, 3 hr	sterilized	365	55.2	1.5	65	45
"	grass, 250°C, 3 hr	sterilized	365	52.7	1.1	93	65
"	oak, 650°C, 3 hr	sterilized	365	78.8	0.6	169	117
"	grass, 650°C, 3 hr	sterilized	365	63.8	0.5	220	153
"	oak, 650°C, 72 hr	sterilized	365	77.3	0.3	329	228
"	oak, 250 °C, 3 hr		365	55.2	2.5	39	27
"	grass, 250°C, 3 hr		365	52.7	1.2	85	59

"	oak, 650°C, 3 hr		365	78.8	0.8	124	86
"	grass, 650°C, 3 hr		365	63.8	1.1	92	64
"	oak, 650°C, 72 hr		365	77.3	0.4	278	193
Zimmerman[4]	Oak, 250°C, 3 hr	sterilized	1173	55.2	3.4	93	65
	grass, 250°C, 3 hr	sterilized	1173	52.7	3.0	104	72
	oak, 650°C, 3 hr	sterilized	1173	78.8	1.1	298	206
	grass, 650°C, 3 hr	sterilized	1173	63.8	0.9	351	243
	oak, 650°C, 72 hr	sterilized	1173	77.3	0.5	658	456
	oak, 250°C, 3 hr		1173	55.2	3.9	81	56
	grass, 250°C, 3 hr		1173	52.7	2.0	156	108
	oak, 650°C, 3 hr		1173	78.8	1.1	279	193
	grass, 650°C, 3 hr		1173	63.8	1.9	166	115
	oak, 650°C, 72 hr		1173	77.3	0.5	587	407
Nguyen et al. 2010	corn, 350°C, slow pyrolysis (BEST Energies)	4°C	365	67.5	9.9	9.6	6.6
"	corn, 350°C, "	60°C	365	67.5	22.2	4.0	2.8
"	corn, 600°C, "	4°C	365	79	5.4	18.0	12.5
"	corn, 600°C, "	60°C	365	79	18.4	4.9	3.4
"	oak, 350°C, "	4°C	365	75.9	3.3	29.8	20.7
"	oak, 350°C, "	60°C	365	75.9	15.4	6.0	4.1
"	oak, 600°C, "	4°C	365	88.4	0.8	124.5	86.3
"	oak, 600°C, "	60°C	365	88.4	13.6	6.8	4.7
Cross and Sohi 2011	sugarcane bagasse, 350°C, 40 min		14	56.2	0.8	4.7	3.3

Table 1. contd....

Table 1. contd....

Source	Biochar Type (feedstock & production conditions)	Incubation Method[1]	Incubation period (d)	Biochar C (weight %)	Biochar C% mineralized[2]	Biochar C MRT[3] (y)	Biochar C Half-life[3] (y)
"	sugarcane bagasse, 550°C, 40 min	35% WHC	14	59.3	0.3	14.7	10.2
Bruun et al. 2012	wheat straw, 525°C, 2 hr		65	69.9	2.9	6.1	4.2
"	wheat straw, 525°C, few seconds		65	49.3	5.5	3.1	2.2

Notes:

1. All incubation at 65% water holding capacity (WHC), in cleaned quartz sand, with microbial and nutrient inoculation added, in dark, at 30°C, unless otherwise indicated.
2. Portion of initial biochar mineralized as measured by analysis of CO_2 efflux.
3. MRT = Carbon mean residence time = 1/apparent degradation rate constant k, $t_{1/2}$ = Carbon half life = $\ln(2) \times$ MRT.
4. These data from an unpublished continuation of incubations reported on in Zimmerman (2010).

most apparent, biochars created at higher temperatures are more resistant to mineralization than biochar made at lower temperatures (e.g., Baldock and Smernik 2002, Hamer et al. 2004, Nguyen and Lehmann 2009, Nguyen et al. 2010, Zimmerman 2010, Cross and Sohi 2011, Bruun et al. 2012). In addition, biochar pyrolyzed for a longer period of time is also more recalcitrant than biochar made of the same feedstock and peak pyrolysis temperature (e.g., Zimmerman 2010, Bruun et al. 2012). In general, biochars made from grasses are more labile than those made from woody material, though this relationship is not as consistent.

Thus, loss of functional groups and aromatization and other chemical changes that occur with increasing pyrolysis temperature and time at elevated temperature, can explain variability in biochar recalcitrance. The volatile matter (VM) content of a charcoal has been suggested as a measure of the degree of thermal alteration or carbonization of a biomass (Antal and Gronli 2003) and high VM biochars have been shown to contain abundant phenolic compounds compared to very few in low VM biochar (Deenik et al. 2010). Though possibly having some adverse effects on plant growth (Deenik et al. 2010, Deenik et al. 2011, Wang et al. 2011, Yang et al. 2011), these compounds have been shown to be available C sources for microbial respiration (Saravanan et al. 2008, Lepik and Tenno 2012, Pradhan et al. 2012). Biochar VM content can be easily approximated according to an ASTM method (D1762–84 1990) as the weight lost when a charcoal is heated in a covered crucible at 950°C for 6 min. As such, it has been suggested that VM might be used as a convenient estimator of short-term biochar lability (Zimmerman 2010, Deenik et al. 2011). Using the set of 28 biochars described in Zimmerman (2010), which include six biomass types and charring temperatures ranging from 250 to 650°C, strong relationships were found between biochar VM content and %biochar C lost over a 1 year incubation with and without microbes (Figs. 1a and 1b, respectively). Similarly strong relationships were also found between %biochar C lost and other interrelated parameters that have been suggested as useful indices of biochar stability such as biochar molar $O:C_{organic}$ ratio (Spokas 2010) and molar $H:C_{organic}$ ratio (IBI 2012).

It is also clear from Table 1 that experimental conditions affect biochar C degradation. Among the factors that have been observed to increase biochar C mineralization in these types of short-term incubations are increases in temperature (Nguyen et al. 2010) and moisture level (Nguyen and Lehmann 2009), and decreases in biochar particle size (Zimmerman 2010). Other factors that are likely important but have not yet been shown to alter mineralization rate include microbial population size and type, oxygen and nutrient level, and biochar/support mineral ratio. Among the greatest influences on the calculated longevity of biochar, however, is the time period over which the incubation is carried out. This can be clearly

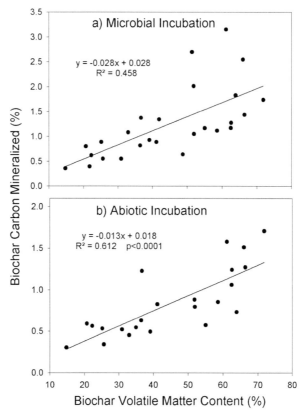

Figure 1. Relationship between biochar volatile matter content and C loss as measured by CO_2 efflux from a) microbial, and b) abiotic incubations. Recalculation of data presented in Zimmerman (2010).

seen by a comparison between the short incubations of Cross and Sohi (2011) and Bruun et al. (2012) which yielded biochar C half-lives ranging from 2 to 10 y (14 and 65 d, respectively), to that of Zimmerman (2010) which yielded half-lives ranging from 27 to 193 y (1 year incubation), to that of a previously unpublished extension of the Zimmerman (2010) incubations to 1173 d which yielded half lives of 56 to 407 y, almost two times greater on average. Thus, selecting an appropriate modeling approach becomes important for accurately predicting biochar C longevity.

2.2 Biochar Mineralization Models

The trend of increasing projected biochar C half-life with increasing incubation period can be explained by the realization that biochar is not a homogeneous material degrading at a single rate of exponential loss (as assumed by the 1G model, Eq. 1). Though the short-term incubation data are well-fit to this model (r^2 values in Table 2), this does not imply ability to predict long-term degradation. The biochar C losses predicted over 100 y range 13–61% (Table 2), do not seem realistic and are much greater suggested by field data (see below).

It is likely that biochar is, instead, a heterogeneous material composed of multiple components that degrade, both biotically and abiotically, at different rates. Thus, another approach commonly used is to model biochar (or soil as a whole) as a mixture of two components (thus, a '2G model'), one which degrades readily (BC_1) and another more recalcitrant fraction (BC_2). These may represent the VM or soluble portion of biochar and the fixed or condensed C fraction, respectively. To apply this approach, the course of CO_2 evolution from biochar can be fit to a double-exponential model using the equation:

$$BC_t = BC_1(1-e^{-k_1 t}) + BC_2(1-e^{-k_2 t})$$ (Eq. 2)

where k_1 and k_2 are the apparent first order degradation rate constants for the labile and refractory pools, respectively, and all biochar C is assumed to be degradable such that $BC_{(total)} = BC_1 + BC_2$. Results of this approach are also well-fit to the data and indicate that a very small portion of the biochars (0–2.4%) degrade very quickly (with half-lives of 1 y or less) and the remainder degrades with half-lives ranging 100's to 1000's of years (Table 2). However, predicted long-term loss rates of 2–59% still may be higher than expected given the longevity of charcoal in soil.

Another modeling approach, utilized by Zimmerman (2010), is one which assumes biochar to be composed of an infinite number of components, with a corresponding continuum of first-order degradation rates (i.e., a time-dependant k). Reactivity continuously and exponentially decreases as the more labile, or perhaps, the more physically accessible organic compounds oxidize or soublize, leaving behind a progressively more refractory or more physically inaccessible residue. The assumption of a wide chemical heterogeneity for biochar is in line with the observation that biochars of different types degrade at different rates (e.g., Zimmerman 2010), are of variable chemical reactivity (Knicker et al. 2007, Knicker et al. 2008) and have organic compound sorption isotherms indicative of heterogeneous substances (e.g., Cornelissen et al. 2006, Kasozi et al. 2010).

The decrease in the reactivity of any organic matter mixture, biochar in this case, has been modeled with a power function (Middleburg 1989) that relates k to time (t), and can be linearized as:

$$\log(k) = -m \log(t) - b$$ (Eq. 3)

with m and b as the slope and intercept, respectively, of the log (k) versus log (t) relationship and can be solved graphically. Table 2 provides k values

Table 2. Comparison of laboratory incubation biochar degradation rate parameters using three different degradation models.

Biochar Type[1]	Abiotic Incubation					Microbial Incubation				
	Grass 250	Grass 650	Oak 250	Oak 650	Oak 650 (72h)	Grass 250	Grass 650	Oak 250	Oak 650	Oak 650 (72h)
1G Model[2]										
k (y^{-1})	4.6E-03	1.9E-03	5.7E-03	2.6E-03	1.0E-03	9.4E-03	5.2E-03	8.8E-03	3.7E-03	1.4E-03
r^2	0.98	0.93	0.97	0.92	0.91	0.95	0.98	0.95	0.95	0.93
C $t_{1/2}$ (y)	150	358	121	266	687	74	133	79	187	484
C loss-10y (%)	4.5	1.9	5.6	2.6	1.0	9.0	5.1	8.4	3.6	1.4
C loss-100y (%)	37.0	17.6	43.6	23.0	9.6	61.0	40.6	58.6	31.0	13.3
2G Model										
k_1 (y^{-1})	1.50	0.56	3.61	4.51	4.33	3.74	0.61	2.58	3.29	6.38
BC_1 (%)	2.1	1.0	0.9	0.3	0.2	2.4	0.7	2.1	0.7	0.2
C $t_{1/2-1}$ (y)	0.5	1.2	0.2	0.2	0.2	0.2	1.2	0.3	0.2	0.1
k_2 (y^{-1})	3.9E-03	2.1E-04	8.9E-03	1.1E-03	8.7E-04	4.4E-03	9.2E-06	6.1E-03	1.7E-03	1.1E-03
C $t_{1/2-2}$ (y)	177	3225	78	613	793	156	74921	114	405	646
R^2	0.99	1.00	0.99	0.98	0.99	0.98	0.98	0.99	0.97	0.99
C loss-10y (%)	5.9	1.2	9.3	1.4	1.1	5.0	2.4	7.9	2.3	1.3
C loss-100y (%)	33.8	3.1	59.2	11.0	8.5	36.3	2.4	46.6	16.3	10.4

Power Model

k-10y (y^{-1})	-1.6E-03	-1.1E-03	-2.4E-03	-7.1E-04	-2.8E-04	-1.5E-03	-7.4E-04	-2.4E-03	-3.2E-04	-3.2E-04
k-100y (y^{-1})	-3.0E-04	-4.2E-04	-6.4E-04	-1.3E-04	-3.8E-05	-2.3E-04	-1.3E-04	-4.5E-04	-3.7E-05	-5.2E-05
r^2	0.91	0.66	0.68	0.95	0.89	0.87	0.77	0.95	0.89	0.79
C $t_{1/2}$ (y)	2.6E+04	2.7E+03	1.7E+03	8.9E+05	4.7E+11	1.6E+05	1.6E+06	5.8E+03	1.0E+07	1.7E+08
C loss-10y (%)	5.7	1.9	5.8	2.8	2.2	7.4	3.3	8.8	5.3	1.6
C loss-100y (%)	10.7	7.3	15.2	5.0	3.0	11.6	5.5	16.5	6.1	2.5

Notes:

1. Biochars used and experimental methods are those described by Zimmerman (2010) but extended to an incubation period of 1173 d. Names refer to biomass and charring temperature (3 h at peak temperature except where 72 h is indicated).

2. The three modeling methods are describes in the text. Parameters listed are k = the apparent degradation rate constant, r^2 = the correlation coefficient for fit of the model to the data, C $t_{1/2}$ = biochar C half-life, C loss-10y and C loss-100y = percent of biochar C model-predicted loss after 10 y and 100 y, respectively. The Levenberg-Marquardt algorithm was used to estimate the value of the model parameters of the 2G model by minimizing the sum-of-the-squared differences between model-calculated and measured values as described in Zimmerman et al. (2011).

and %C loss calculations after 10 and 100 y predicted by this model. Losses of 5 to 17% after 100 y and half-lives ranging from thousands to hundreds of millions of years may be more reasonable given that charcoal fragments are found in ancient soils and sediments (e.g., Scott 2000, Dodson et al. 2005). While this model may be the most intellectually satisfying, it has the drawbacks of being more difficult to apply and requiring a longer period of data collection.

There are, however, major shortcomings to the laboratory incubation approach to estimation of long term biochar degradation which cannot be overcome simply by developing an appropriate degradation model. One is that a closed laboratory chamber can never truly simulate the open system that is the soil environment. Unaccounted for stimulatory effects on biochar degradation include UV exposure, rainwater infiltration, bioturbation, positive priming by soil OM and possible inhibitory/protective effect include mineral interaction and soil OM adsorption (see later discussion). Even the very variability in climate parameters such as temperature, freeze/thaw cycles and saturation/desaturation have been shown to stimulate microbial degradation (Sun et al. 2002, Cravo-Laureau et al. 2011). Oak biochar was found to mineralize most rapidly under alternating saturated–unsaturated conditions (Nguyen and Lehmann 2009). Another issue is the possible decrease in microbial biomass over incubation time due to non-ideal conditions such as nutrient limitation and buildup of metabolic products (Spokas 2010, and references therein). And laboratory incubations may not include either fungi or conditions conducive to fungi growth, which may be major players in biochar biodegradation (Lehmann et al. 2011). Thus, it is important to compare laboratory degradation data to that of field studies of different types.

2.3 Field Studies

A limited number of studies have been carried out using measurements of BC distributions in soil or sediment profiles with known accumulation rates and pyrogenic OM input histories to calculate apparent BC loss rates in the field. Their results range from showing no loss of pyrogenic OM at all to BC half-lives of 3.3 y (Table 3). Together, they show losses of biochar, to be substantial under some environmental conditions and negligible in others. The range may be due to the fact that each of these studies utilizes different BC materials, different analytical techniques and different soil and climate types. Only one study (Major et al. 2010) followed losses in the field using biochar of known starting composition and amount. The remainder relied upon BC produced by natural fires or slash and burn agriculture. These studies can yield results more nuanced than is indicated by a simple reading of a table such as this. For example, the chronosequences examined

Table 3. Biochar C loss rates derived from field studies.

Source	Location/Soil/BC/ Experimental	BC quantification method	Time period (y)	Biochar C loss (%)	Biochar C $t_{1/2}$
Major et al. 2010	Colombia savanna Oxisol, mango prunings, charred in kiln, 400–600°C, 48 h	natural abundance d ^{13}C-OM	2	3.3	42
Nguyen et al. 2008	Western Kenya, Humic Nitosols,	^{13}C-CPMAS NMR	30	70	25
	forest slash and burn sites of different ages with no subsequent burning	hand picking	5	84	3
Hammes et al. 2008	Russian Chernozem sampled before burning cessation and 100 years later	benzene polycarboxylic acids (BPCAs)	97	24	245
	"		104	20	323
Schneider et al. 2011	Western Kenya, forest to ag. converted by slash and burn 100 y ago and less	BPCA	100	0	—
Vasilyeva et al. 2011	Russian Chernozem soil, cessation of burning (steppe and fallow)	BPCA	55	0	—
Bird et al. 1999	Zimbabwe sandy savanna soil, natural burning cessation	Acid-dichromate oxidation	51	47	56
Middleburg et al. 1999	Abyssal plain marine sediment exposed to oxygen	Thermal oxidation (soot only)	10–20 ky	63	7–14 ky

by Nguyen et al. (2008) after forest clearance by fire at different times, revealed an initial 30 y phase of rapid BC loss (or 5 y of visible charcoal loss) followed by at least 70 y of stable concentrations. One might conclude that this shows that a portion of the BC degrades on decadal timescales, while the residual portion is lost at much greater than centennial scales.

However, these types of studies are associated with their own set of uncertainties. Among them are: 1) the possibility that BC has been lost to erosion or vertical transfer with the soil profile, 2) reliance upon either the assumption that BC production in the natural environment was constant or ceased altogether at a known time, and 3) uncertainties associated with the quantification of charcoal or BC. Even if the assumptions of 1) and 2) can be satisfied, the last of these is still of much concern. Each technique only quantifies a portion of the pyrogenic C present. Thermal oxidation, for example, is now known to preferentially detect soot C (Hammes et al. 2007) and the polycondensed aromatic domains detected by ^{13}C-NMR spectroscopy have been shown to represent only a small portion of pyrogenic material in some cases (Sharma et al. 2004, Knicker et al. 2005). Further, it is not known that isotopic signatures, the proportion of condensed aromatic C as detected by ^{13}C-NMR techniques, or the content of the biomarker benzene polycarboxylic acids (BPCAs) of BC remain constant as BC degrades. They most likely do not. For example, both the BCPA (Schneider et al. 2010) and spectroscopy preferentially detect the highly condensed portion of char more likely to be produced at higher temperatures and will be preferentially retained over time. And aged biochar was found to leach ~50 times more BCPA's than fresh biochar (Abiven et al. 2011). Thus, much as laboratory incubation experiments are most likely to overestimate the rate at which biochar is degraded by focusing attention on the most labile portion degraded early, field studies are likely to underestimate biochar degradation with their bias toward following the most recalcitrant portion of biochar. Additionally, laboratory incubations using CO_2 efflux measurements underestimate true environmental stability by neglecting loss mechanisms other than microbial or abiotic oxidation while field measurements include the effects of, but do not necessarily properly evaluate, all types of BC loss.

3 Mechanisms of Black Carbon and Biochar Loss

Of the possible mechanisms for biochar loss from soils, degradation/ mineralization (abiotic and biotic), erosion or downward translocation, leaching/solubilization, volatilization and consumption by later fires, only the first has been the subject of considerable research as yet. Aspects of each of these mechanisms are discussed in the following.

3.1 Biotic Black Carbon Degradation

The evolution of CO_2 from incubations of biochar are often attributed to microbial respiration of biochar when, in actuality, they are the net product of both abiotic and biotic processes. There is a good deal of evidence for the stimulatory effect of microbial processes on biochar degradation. Biological utilization of BC was first observed by Potter (1908) by recording CO_2 evolution from 20-day charred wood and coal incubated with a *Diplococcus* culture. By comparison, Potter saw no CO_2 evolution from either sterilized or dry incubations, confirming the involvement of microbes. In other early work, Shneour (1966) measured greater release of CO_2 from a ^{14}C-labeled charcoal than in parallel sterilized incubations. A longer term study (1 y) using a range of biochars showed up to twice the CO_2 release from incubations inoculated with a consortia of soil microbes (Zimmerman 2010) compared to sterilized incubations. The proportion of biochar C mineralized in these experiments that could be attributed to microbial involvement was greatest for chars produced at lower temperatures, thus richer in organic acids and phenols, suggesting a microbial preference for specific biochar OM components.

Measurements of increased microbial biomass via plate counting, DNA extraction, fumigation or phospholipid fatty acid extraction have also provided evidence of microbial response to pyrogenic OM additions. A number of researchers have also found increases in microbial biomass, along with basal respiration rate, that directly corresponded with increasing quantity of charcoal added to soil (Steiner et al. 2008, Kolb et al. 2009, Steinbeiss et al. 2009). The same result was found for amendments of pyroligneous acid, a condensate of the smoke produced during biomass combustion (Steiner et al. 2008). However, most studies of these types have not distinguished between increases in microbial biomass due to utilization of pyrogenic OM as substrates for microbial growth and activity versus that which may be due to the ability of charcoal, with its high surface area and porosity, to enhance the microbial soil environment by maintaining favorable moisture and aeration conditions and protection from grazers. The one study that has shown direct incorporation of pyrogenic C by microbes found that 1.5–2.6% of a ^{14}C-labeled biochar produced from rye grass shoot litter at 400°C (13 h) was incorporated into microbial biomass after 624 d of incubation with soil (Kuzyakov et al. 2009).

There is also uncertainty as to the specific microbial organisms responsible for pyrogenic OM degradation. Likely candidates include fungi, Proteobacteria (O'Neill et al. 2009, Khodadad et al. 2011) and Actinobacteria (Baath et al. 1995, Khodadad et al. 2011). An important role of fungi in biochar degradation is widely hypothesized due to their known ability decompose lignin, low-grade coals and other refractory OM (Scott et al. 1986, Gotz and Fakoussa 1999, Hofrichter et al. 1999, Wengel et al. 2006) and the often observed positive association between soil biochar and fungal mycorrhiza (reviewed in Warnock et al. 2007). The many effects of biochar on soil biota have been reviewed in detail elsewhere (Thies and Rillig 2009, Lehmann et al. 2011).

3.2 Abiotic Oxidative Black Carbon Degradation

Oxidation of biochar, both on its surface and in bulk, clearly occurs over time. Rates of CO_2 production and O_2 consumption were strongly correlated in incubations of a variety of biochars (Spokas and Reicosky 2009). Boehm titration, Fourier-transform infrared (FT-IR) spectroscopy, and X-ray photoelectron spectroscopy (XPS) have shown increases in oxygen content and phenolic and carboxylic functional groups to be initiated at biochar's surface both in 4-month laboratory incubations and over decades in the soil (Cheng et al. 2006, Cheng et al. 2008, respectively). Studies using ^{13}C-NMR appear to indicate that oxidation may penetrate beyond biochar's surface with bulk aryl-C losses of up to 53% and carboxyl/carbonyl C and O-aryl C content increased to 29% over a twenty month incubation of grass biochar produced at 350°C (Hilscher and Knicker 2011a).

Attribution of biochar degradation or a portion of the observed degradation to purely abiotic causes is more difficult. While some studies report no or little loss of biochar C in the absence of microbes (Potter 1908, Shneour 1966, Santos et al. 2012), a number of studies suggest that abiotic processes play an important, perhaps even dominant, role in degrading biochar. Zimmerman (2010), for example, reported that, over 1 y, sterilized incubations of a range of biochars released 50–90% of the CO_2 released by inoculated ones. While it can be argued that abiotic conditions are difficult to maintain during longer-term incubation experiments (Knicker 2011), the different pattern of microbial versus abiotic degradation observed suggests real differences in degradation mechanism between biotic and abiotic ones. In another study, addition of microbes did not alter the extent of oxidation of biochar surfaces during 4-month laboratory incubations (Cheng et al. 2006). And the lack of a lag phase at the start of charcoal incubations has been suggested as evidence for the predominance of non-microbial degradative processes, at least during early phases of degradation (Bruun et al. 2008). Finally, Nguyen et al. (2010) showed that biochar degradation rate increased as temperature was raised from 4 to 60°C, unlike what would be expected of microbial processes which are usually maximal closer to 30°C.

Most likely, both abiotic and microbial oxidative processes are important and, to some extent, interdependent. Chemical studies suggest a two part process in which aryl C structures are first abiotically oxidized, perhaps via oxygen chemisorptions reactions assisted by the high density of electron-donating pi-bonds (Contescu et al. 1998), or hydrolyzed to form catechol-like structures. The O-aryl rings can then be cleaved or converted to C-containing functional groups (Hilscher et al. 2009, Hilscher and Knicker 2011a). These organic components are now both more labile and more water-soluble, thus more accessible to microbes for use as respiratory substrates. Degradation experiments comparing fresh and biochars abiotically aged by oxygen, water, sunlight etc., in parallel and in series, should be carried out to explore the relationships between these processes.

3.3 Non-Oxidative Abiotic Black Carbon Loss

Dissolution of inorganic C minerals or amorphous phases, desorption of CO_2 or volatilization of organic compounds have all been proposed as possible biochar C loss mechanisms. While each of these processes would be expected to be short-lived (days to years?), they need to be better understood and accounted for when carrying out degradation experiments or C budget calculations.

Using X-ray diffraction spectra, a number of studies have reported the presence of calcite and other carbonate phases in biochar made by pyrolysis between 300 and 500°C (Cao and Harris 2010, Singh et al. 2010a, Inyang et al. 2011, Yuan et al. 2011). Calcite is not expected to be present in biochars made at higher temperatures as thermal decomposition of calcite begins to occur at about 600°C and is completed at 850°C (Rodriguez-Navarro et al. 2009). By measuring the release of ^{14}C during acidification of biochar made from ^{14}C-labeled barley root at 375°C, Bruun et al. (2008) calculated that carbonate represented up to 11% of the biochar C. They attributed the C released during early phases of incubation of this material wholly to this source. In another study, Jones et al. (2011) observed a 50% decrease in CO_2 released from biochar incubations in soil after rinsing with water and a 5-fold decrease when biochar was pre-treated with acid. They concluded that abiotic release of mineral carbonate C contained in the biochar was a major factor but did not report having carried out measurements of inorganic C.

The set of 20 biochars (pyrolyzed at 250–650°C) reported on in Zimmerman (2010) was tested for inorganic C content by acidifying with 10% phosphoric acid and measuring evolved CO_2 by coulometry (UIC Inc., Joliet, IL). After 5 min acidification, $CaCO_3$-C% (by weight) release ranged 0–0.5% and after 72 h acidification, 0.01–0.6%. Greatest carbonate contents occurred in high temperature (650°C) biochars and in grasses. Thus, we calculate that, while inorganic C could have accounted for <3.7% of the CO_2 released from low temperature chars (produced at 250 or 400°C) during incubation, it could have accounted for as much as 42% (26% on average) of the C released from chars made at 525 of 650°C. However, the amount was probably far less given that strong acid was required for all of this carbonate C to be released. However, this should serve as warning that inorganic C should be considered during future degradation incubations.

Thus far, there have only been qualitative analyses of the volatile compounds able to be desorbed from biochar. Spokas et al. (2011) identified over 140 individual compounds released using headspace thermal desorption at 150°C coupled to capillary gas chromatographic-mass spectrometry. Lower temperature biochars produced the greatest amount and number of compounds and CO_2 and other short-chain alkanes and alcohols were always among the volatile compounds detected. In only one indication of the amount of biochar C that may be lost to volatilization, Spokas et al. (2009) found that pre-treating biochar under a vacuum reduced CO_2 efflux in incubation by 7%. Further research should be carried out in this area.

3.4 Black Carbon Loss by Leaching

A number of recent studies have examined losses of C, nitrogen and phosphorous as aqueous leacheate from biochar-amended soils driven by interest in biochar's effect on plant-available nutrients. However, the quantification of leached dissolved organic carbon (DOC) from biochar itself is of interest from the standpoint of biochar's environmental longevity. The fate of that biochar-derived DOC is also of importance in establishing biochar's effect on soil C balance and global C cycling.

The production of colored leacheate when pyrogenic OM is placed in water has long been noticed. Losses of C by this process may be significant as researchers have identified substantial amounts of pyrogenic substances in rivers and the ocean using BC analyses, molecular markers and ultrahigh resolution spectroscopic tools (Mitra et al. 2002, Mannino and Harvey 2004, Hockaday et al. 2006, Hockaday et al. 2007). This dissolved pyrogenic OM is identified as condensed aromatic ring structures of relatively low molecular weight (<1000 Da), and extensively substituted with oxygen-containing functional groups (Kim et al. 2004, Dittmar 2008). However, it should be understood that the similarity of these water-soluble condensed aromatics in natural waters to the degradative products of BC does not confirm their common origin. On the other hand, there may be many more pyrogenic compounds in natural waters not yet recognized as such.

Only a few studies have quantified the generally low amounts of C lost by biochar using laboratory aqueous leaching experiments. A mixed hardwood-derived biochar pyrolysed at 450°C for 48 h released only 0.04% of its total C after repeated rinsing with distilled water (Jones et al. 2011). And a 100 y-old charcoal yielded 0.2% of its C when extracted at pH 4.4 with HCl after 48 h (Hockaday et al. 2006). Abiven et al. (2011) extracted a fresh chestnut wood biochar (450°C for 5 h under N_2) and 10 y aged natural charcoal using a more extensive routine (8 g bulk dry in 100 ml, 6 h repeated 6 times). They found that the fresh and aged biochars released 0.2% and 0.1% of their total C, respectively, in the combined soluble and colloidal fractions. Though these losses are small, the similarity between C leached from fresh and aged biochar, and our unpublished results (below) indicate that losses vary considerably with biochar type and can be more substantial over longer periods of leaching.

We tried to estimate long-term desorbable biochar C by carrying out both multiple consecutive batch and continuous flow-through column leaching experiments using the set of biochars described in Zimmerman (2010). For batch leaching, 20 g of each biochar was added to 400 mL distilled deionized (DI) water in 500 ml plastic bottles and placed horizontally on a mechanical platform shaker (150 rpm) in the dark. Weekly, over 3 months, the bottles were centrifuged (4500 rpm) and the supernatant was carefully removed via pipette. The remaining sample was weighed to determine the amount of entrained solution present before addition of 400 mL more DI water. The DOC in the supernatant was measured by TOC analyzer (catalyzed high temperature combustion) and leached DOC was calculated after taking into account the entrained DOC left in each new round of leaching. Flow-through column desorption was carried out using a packed chamber filled with 10 g biochar plumbed with water flowing upward for full contact and driven by a column of water of a height adjusted to achieve the desired flow rate of about 150 ml d^{-1}.

The results point to a number of interesting trends in biochar C leaching (Fig. 2). First, after nine rounds of leaching, biochar C losses ranged from 0.1 to 1.3% for all biochars tested. As expected from their greater VM content, biochars created at lower temperatures lost the greatest proportion of C via leaching. Second, biochar C leachability varied with parent biomass type following the trend pine wood ≈ oak wood < grass. The much lower C losses measured previously by others can be attributed to their examination of low VM biochars or to their use of only a single leaching period. Lastly, batch and column C leach rates were similar after normalizing for amount of biochar and volume of extractant (water). We have conducted a number of other leaching experiments using the same materials but different solid/solution ratios, contact times, flow rates, etc., which all have yielded similar normalized C yields (Mukherjee and

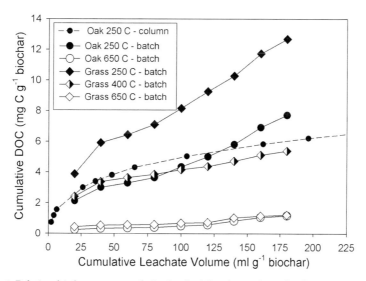

Figure 2. Relationship between cumulative leached C and cumulative leachate volume during successive batch (solid lines) and continuous flow column (dashed line) DI water extractions of biochars made from Laural Oak wood and Gamma grass combusted at 250°C (in oven with atmosphere) and 400 and 650°C (under flowing N$_2$) as described in Zimmerman (2010).

Zimmerman. In press). Further indication that biochar C loss by leaching may be significant, though column leaching experiments were carried out to a total leach volume of 7.4 L (almost four times that depicted in Fig. 2), we never observed a decrease in C leach rate. Similar leaching experiments carried out on these same biochars which had been aged for 1 y outside in north Florida, during which time 109 cm of rain fell (fully described in Mukherjee 2011), yielded similar, even greater leach rates in some cases than non-aged biochars. We calculate that, assuming these laboratory-observed leaching rates remain constant, losses of biochar C by leaching over 1 y of average north Florida rainfall would be ~1% and 2% for high and low temperature biochars, respectively. These C loss rates are similar to those measured via laboratory CO_2 efflux experiments during 1 y and could be modeled similarly. However, if leaching C losses remain constant over time rather than decrease as do mineralization rates, it could add up to much greater long-term C losses.

There are two likely explanations for these observations of seemingly continuous leaching of biochar C. First, desorption of organic compounds from biochar may be driven by a chemical equilibrium at the solid-solution interface. Thus, movement of chemisorbed organic compounds such as organic acids and phenols between biochar's surface and the solution, which likely occurs in the confined space of biochars pores, would progress incrementally with each change of solution chemistry via reversible reactions that could be described by thermodynamic equations. On the other hand, hydrophobic compound sorption-desorption is more typically described using partitioning theory, which would liken the situation to volatile molecules permeating a more stable organic framework. Certainly, more work is needed in this area.

A second explanation for continuous leaching is that, over time, abiotic or biological degradation increases the solubility of organic compounds on biochar's surface by adding polar functional groups to aromatic units. For example, oxidation of soot has been report to produce water-soluble organic compounds (Decesari et al. 2002, Kamegawa et al. 2002). Similarly, Reiss (1992) found that fungi was able to solubilize up to 78% of coal, by mass, and this ability varied directly with its VM and oxygen content. Knicker (2011) pointed out that this increase in biochar solubility would be expected to occur when carboxyl groups are deprotonated, which would occur more readily under the alkaline conditions that are typically produced by biochar amendments to soil (e.g., Atkinson et al. 2010).

But the effect of biochar solubilization on the C content of a biochar-amended or natural fire-impacted soil hinges on the fate of these leached organic compounds, which could include 1) percolation through the soil and export out of the watershed via rivers and groundwater, 2) sorption onto soil OM or minerals and long-term retention, or 3) mineralization through abiotic or microbial degradation. A case could be made for the occurrence of each of these, depending upon biochar and soil type, climate etc. Claims of detection of abundant products of BC degradation in natural waters would argue for persistence and mobility through soil. However, in the few studies where C leaching from biochar-amended soils has been measured, very little dissolved pyrogenic C was detected. For example, Major et al. (2010) only measured biochar C losses of ~0.02% in percolating water collected by soil lysimeters over almost 2 y from a biochar-amended Oxisol.

Some biochar-amended soils show increased DOC fluxes over non-amended controls (e.g., Bell and Worrall 2011, Mukherjee 2011), which may also be due to increases in plant or microbial productivity, while others do not (Novak et al. 2010). No biochar C was detected in the K_2SO_4-extracted DOC of a ^{14}C-labeled ryegrass biochar after 624 d incubation with soil. In laboratory column experiments with grass biochar-amended soils, we found evidence of 20–40% of biochar-C leachate sorption by a sandy Entisol, but its full mobility through a clayey Utilsol (Mukherjee 2011). Sorption of aromatic phenols and carboxylic acids to soil minerals (Huang et al. 1977, Kaiser and Guggenberger 2000, Hyun and Lee 2004) and to biochar-itself (Kasozi et al. 2010) has been shown experimentally. However, the extents to which these sorbed components may be later desorbed or are microbially available is an area for future research.

Abiotic or microbial mineralization is another likely fate of leached dissolved biochar C. Soil exposed to the water soluble portions of wood smoke had enhanced rates of CO_2 respiration and O_2 consumption (Focht 1999, Steiner et al. 2008). We carried out *in vitro* abiotic (autoclaved) and microbial (inoculated with a consortia of soil-extracted microbes and nutrients) incubations of biochar leachates. For each biochar type, 35 ml leachate adjusted to a DOC concentration of 25 mg L^{-1} was incubated in the dark at 30°C over 313 d and evolved CO_2 was measured monthly. We found the portion of biochar C mineralized in abiotic incubations to range 10–61% (35% on average) and in microbial inoculated incubations, 75–37% (56% on average) (Fig. 3). Thus, on average, 61% of the mineralization observed could be attributed to abiotic processes, assuming these incubations remained sterile. There was no consistent correspondence between biochar type and mineralizability of leachate. However, the highest temperature biochars produced somewhat more refractory leachate.

In summary, it is likely that, both in laboratory and in the field studies, the amount of biochar C measured as mineralized by CO_2 efflux is controlled by processes of OM desorption and mineralization of dissolved BC because 1) the quantity of biochar-C mineralized (~1–2% over a year) is roughly equivalent to the amount likely to be leached, and 2) correlations between the quantity of biochar-C mineralized and biochar VM contents or other proxies of potential solubilization. Between mineralization and sorptive

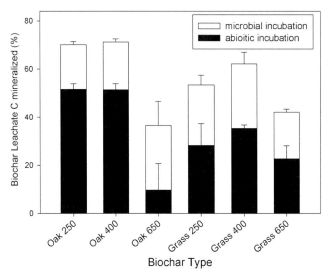

Figure 3. Portion of C in biochar leachates mineralized to CO_2 after 313 d sterile and microbial inoculated incubation. Error bars represent standard deviations calculated from duplicate treatments. Biochar types are the same as shown in Fig. 2 and described in Zimmerman (2010).

retention of dissolved BC, it seems there would be little pyrogenic C export from the soil. This needs to be reconciled with the apparent contradiction of observations of large amounts of dissolved BC in natural waters.

3.5 Black Carbon Losses by Erosion/Translocation

While there have been few direct quantifications of losses of particulate soil BC or biochar by horizontal movement (erosion) or vertical movement (downward translocation), there is quite a bit of indirect evidence that significant losses, perhaps even the majority of losses of soil BC, may occur by these processes. First, erosion is implied by the observation that BC makes up a large portion of the carbon in bogs and streams (Guggenberger et al. 2008) and riverine and marine sediment (2–38%, Masiello 2004). Strengthening the case for a soil BC source for this material, studies have found temporal correspondence between decadal-scale periods of regional drought and coastal deposits enriched in BC (Bird and Cali 1998, Pederson et al. 2005, Mitra et al. 2009).

Second, erosion of soil containing BC should be expected given that natural fires may denude the landscape of erosion-deterring vegetation and biochar-enriched soils may have been preferential sites of high intensity agriculture. Using experimental plots and a rainfall simulator, Rumpel at al. (2006) found that soil erosion was most severe in an area under intensive slash and burn practice. And even within a specific soil, BC particles may be preferentially exported compared to other soil OM due to their predominantly small size and low density (Skjemstad et al. 1999). Rumpel et al. (2009) found that 7–55% of pyrogenic C added to the soil via a surficial grass fire was lost to erosion and another 23–46% was moved vertically to greater soil depths.

One might expect erosional losses of biochar to much less in level agricultural systems in which biochar is initially well-mixed into soil. However, downward movement of biochar, either as discrete particles, colloids or in dissolved form, may still be significant. Dai et al. (2005) found no effect of controlled burns on the BC content of a north Texas soil in the 0–10 cm depth interval, but a significant increase at the 10–20 cm interval suggesting downward movement of BC. Though homogenized into soil at the start of a laboratory column experiment, Hilscher et al. (2011b) found downward movement of 2.3% of the ^{13}C-labeled biochar after 28 months and attributed this to leaching and re-adsorption of pyrogenic C or to translocation of clay-sized particles. However, in the one field study that attempted to follow all BC losses from a biochar-amended soil in the field, leaching, downward movement and mineralization only accounted for losses of <3% leading them to conclude that 20–53% of the applied BC must have been lost by surface erosion (Major et al. 2010). Clearly, losses of these types will vary with such factors as climate, topography, soil type and biochar application routine.

3.6 Black Carbon Losses by Later Fires

Indirect evidence for losses of pyrogenic C by consumption during later fires may been found in the observation that, in a number of studies, no additional soil BC was registered after multiple burning events compared to single events (Dai et al. 2005, Knicker et al. 2006). Also, studies have calculated BC stocks and ages much less than would be expected based on estimates of landscape fire frequency over long time periods (e.g., Czimczik et al. 2003, Ohlson et al. 2009). Of course, these losses could be due to other causes such as mineralization and erosion. Direct evidence for biochar losses by repeat fires is lacking.

4 Biochar-Soil C Interaction and Stability

While it appears that the greatest biochar C losses occur by microbial and abiotic oxidation (mineralization), it is not clear the extent to which this will be impeded or stimulated by the interaction of biochar with soil. For example, a few studies have found that some portion of soil BC can be found embedded within microaggregates (Glaser et al. 2000, Brodowski et al. 2006) which could slow degradation by limiting access by oxygen or microbial enzymes. Soil OM, being a more labile C source than pyrogenic C, could stimulate the production of microbial exoenzymes, leading to the co-metabolism of BC. This accelerated mineralization of a refractory OM component when stimulated by the presence of a labile C source has been termed 'positive priming' (Kuzyakov et al. 2000). Making matters more complicated, the presence of biochar could either impede or accelerate the mineralization of soil OM. For example, given that biochar is both porous in nature and has high affinity for natural OM (Kasozi et al. 2010), it could be expected to sequester non-BC soil OM within its pore network and protect it from degradation by oxidants and exoenzymes. Or biochar may sorb soil nutrients, limiting microbial activity. Alternatively, BC may accelerate soil C mineralization by providing nutrients or a habitat favoring increased microbial heterotrophic activity (Thies and Rillig 2009). Another idea is that following biochar-induced increased plant growth, microbial activity stimulation occurs in the rhizosphere (Graber et al. 2010).

At this point, a great number of studies have measured changes in CO_2 efflux following biochar amendment to soil. Generally, the results find either increased soil CO_2 efflux due to biochar addition (Kolb et al. 2009, Kuzyakov et al. 2009, Spokas and Reicosky 2009, Major et al. 2010, Zimmerman et al. 2011, Jones et al. 2012) or no significant change in soil CO_2 efflux (Kuzyakov et al. 2009, Spokas and Reicosky 2009, Novak et al. 2010, Singh et al. 2010b, Cross and Sohi 2011, Karhu et al. 2011, Case et al. 2012). One field litterbag study found greater losses of humus in the presence of charcoal than in its absence, possibly indicating positive priming (Wardle et al. 2008). All these results should be viewed with the understanding that they do not indicate positive priming or lack of priming of soil OM mineralization by biochar (or vice versa) if they do not correct for the increased C mineralization that would be due to biochar alone (from abiotic or microbial oxidation, or release of inorganic C in biochar).

A better understanding of the effects of the complex and multiple possible interactions between biochar and native soil OM can only be had by experimental methods which allow for the quantification of biochar C mineralization and native soil C mineralization separately. Results of studies that have done this are assembled in Table 4. In general, they show biochar C degradation rates similar to or greater than those carried out on biochar alone, ranging from losses of 0 to 9%, but with a median values of 1 to 2%. Further, the change in soil C mineralization rate observed with the addition of biochar ranged from +150 to –87%, with a median value of +6%, or almost no change. Clearly, the direction of priming is highly variable and seems dependent upon biochar and soil type and time. Likely due to their greater lability, low temperature biochars generally have been found to

Table 4. Biochar C and soil C loss rates derived from combined soil/biochar incubation studies.

Source	Biochar feedstock and production conditions	Soil	Incubation and Quant. Method[1]	Incubation period (d)	Biochar %C mineralized	Biochar C $t_{1/2}$ (y)	Soil C min. % change[2]
Kuzyakov et al. 2009	^{14}C-labeled ryegrass, 400°C, 13 h	Haplic Luvisol	^{14}C-CO$_2$/ scint. count	1181	4.0	55	0
"	"	loess	"	"	4.0	55	decrease
Steinbeiss et al. 2009	^{13}C-labeled hydrothermal pyrolysis-glucose	arable Eutric Fluvisol	20–25°C	120	8.0	2.7	53
"	^{13}C-labeled hy py yeast	"	"	"	9.0	2.4	150
"	^{13}C-labeled hy py glucose	forest Cambisol	"	"	3.0	7.5	57
"	^{13}C-labeled hy py yeast	"	"	"	6.0	3.7	108
Major et al. 2010	mango prunings charred in kiln, 400–600°C, 48 h	Colombia savanna Oxisol	static chamber CO$_2$ measured in field	730	2.2	62	25.5
Zimmerman et al. 2011	grass, 250°C under N$_2$	forest Alfisol	50% WHC	90	2.9	5.8	61.3
"	grass, 650°C under N$_2$	"	"	"	0.9	18.3	–15.6
"	grass, 250 °C under N$_2$	wetland Mollisol	"	"	6.8	2.4	0.1
"	grass, 650°C under N$_2$	"	"	"	0.2	114	–76.4
"	grass, 250°C under N$_2$	forest Alfisol	"	90–500	1.9	41	–75.1
"	grass, 650°C under N$_2$	"	"	"	0.1	613	–87.1

Table 4. contd....

Table 4. contd....

Source	Biochar feedstock and production conditions	Soil	Incubation and Quant. Method[1]	Incubation period (d)	Biochar %C mineralized	Biochar C $t_{1/2}$ (y)	Soil C min. % change[2]
"	grass, 250°C under N_2	wetland Mollisol	"	"	3.8	20	–9.9
"	grass, 650°C under N_2	"	"	"	2.6	29	–49.8
Cross and Sohi 2011	sugarcane bagasse, 350°C, 40 min	fallow silty-clay loam		14	1.1	2.5	15.1
"	" 550°C, 40 min	"		"	0.1	29	18.0
"	" 350°C, 40 min	arable loam		"	0.7	3.6	25.1
"	" 550°C, 40 min			"	0.0	0	6.1
"	" 350°C, 40 min	grassland loam		"	0.4	7.1	–9.3
"	" 550°C, 40 min	"		"	0	0	–36.0
Keith et al. 2011	^{13}C-depleted Eucalyptus, 450°C, 40 min pyrolysis	Vertisol		120	0.75	30	7.4
"	" 550°C, 40 min pyrolysis			120	0.4	57	19.4
Santos et al. 2012	^{13}C-enriched pine, 450°C, 5 h under N_2	forest - andesitic parent	55% WHC, 25°C	180	0.37	92	–10
"	"	forest - granitic parent	"	"	0.41	83	0

Luo et al. 2011	straw, Ar-flushed pyrolysis, 350°C, 30 min	Aquic Paleudalf, low pH	40% WHC, 25°C	87	0.61	27	304
"	"	" high pH	"	"	0.84	28	202
"	" 700°C, 30 min	Aquic Paleudalf, low pH	"	"	0.14	118	136
"	"	" high pH	"	"	0.18	92	74

Notes:

1. All incubation at 60–65% water holding capacity (WHC) in dark at 30°C, unless otherwise indicated. The amount and proportion of biochar C and soil C mineralized was measured by analysis of natural abundance of d^{13}C-CO$_2$ efflux, unless otherwise indicated.
2. Change in soil C mineralized or mineralization rate with, relative to without, added biochar.

stimulate soil OM mineralization to a greater extent than high temperature ones (Cross and Sohi 2011, Zimmerman et al. 2011). The effect of soil type is more difficult to discern. But soil C mineralization seemed to be inhibited by biochar to a greater extent in soils of lower organic C content (Kuzyakov et al. 2009, Zimmerman et al. 2011), suggesting that protective sorption of soil OM onto biochar is limited in extent.

As with biochar-alone incubations, time period of observation seems an important factor. For example, Luo et al. (2011) demonstrated the initiation, progress and termination of a positive priming effect by biochar on native soil C over the course of 3 months. Zimmerman et al. (2011) found that priming effect swung from positive to negative after a period of about 3 months (Table 4) and longer studies such as that of Kuzyakov et al. (2009) and Santos et al. (2012) tend to show a low or negative effect of biochar on soil OM mineralization. Finally, Liang et al. (2010) found that mineralization of native soil C was 64–82% lower in the BC-rich Anthrosols compared to nearby BC-poor soils. Together, these studies suggest that, following an initial stage of positive priming in which soil C mineralization may be stimulated (3 months?), an inhibitory effect of biochar on soil C mineralization takes over as soil OM is sorbed and perhaps encapsulated within biochar-OM mineral microaggregates. Organic matter sorption onto biochar surfaces has been shown to be kinetically limited by slow diffusion into the subnanometer-sized pores dominating biochar surfaces (Kasozi et al. 2010). This effect should be greater for higher temperature chars which possess greater porosity. These time-, soil- and biochar type-dependant trends explain the wide range of experimental degradation results obtained in various biochar-amendment studies. Some of the concepts described here for C, likely also apply, to some degree, to biochar-soil nutrient dynamics such as for N and P, particularly in organic forms.

5 Mechanisms of Biochar Stabilization

There have been a number of recently published reviews discussing the mechanisms which may stabilize BC or biochar in soil and account for biochar's longevity in the environment (Czimczik and Masiello 2007, Lehmann et al. 2009, Knicker 2011). Certainly inherent chemical or molecular-recalcitrance is a major factor, though this is highly dependent upon environmental conditions, as discussed above. There has been much study of whether BC is physically stabilized by its interaction with soil minerals, soil OM, or occlusion within microaggregates that include both OM and clay-sized minerals. For example, Liang et al. (2008) found that BC-rich Anthrosols had a high proportion (72–90%) of its C in the organo-mineral fraction. Using X-ray spectromicroscopy, discrete particles of BC were observed embedded within intact microaggregates (Lehmann et al.

2008). And Al, Si, and to some extent Fe, were found associated with BC particle surfaces within 30 y after deposition in soil (Nguyen et al. 2008). However, Brodowski (2006) found soil BC to be widely distributed across a range of density fractions including microaggregates and discrete particles and major portions of the BC in soil can be found within the lightest soil fraction (Glaser et al. 2000, Rumpel et al. 2006, Murage et al. 2007). In addition, across a range of North American prairie soils, no correlation was found between BC content and soil properties including clay content (Glaser and Amelung 2003).

Though the importance of mineral-association to biochar preservation is debatable, soil C and soil BC abundances are usually correlated (Glaser and Amelung 2003, Czimczik and Masiello 2007, Vasilyeva et al. 2011) indicating that BC and native soil OM are preserved by similar processes. For example, biochar particles or leachates and native OM may be stabilized separately or held together via ion bridging with Ca^{2+}, or with Al and Fe oxyhydroxides (Glaser et al. 2000, Czimczik and Masiello 2007). These processes would likely grow in effectiveness with the surface oxidation that occurs with biochar aging. This may partially explain why aged soils enriched in BC such as *terra preta* are also enriched in non-BC OM.

Organic matter sorption onto surfaces as well as into mineral nanopores is an OM protection mechanism that has received wide attention (e.g., Keil et al. 1994, Kaiser and Guggenberger 2000, Mayer et al. 2004). Thus, OM sorption onto biochar's surface and within its pores may be an effective mutual preservation mechanism. While microbes may be excluded from micron-sized pores (those that are typically seen on SEM images of biochar particles), we have found that the majority of biochar surface area actually exists within nanometer diameter pores (Kasozi et al. 2010, Mukherjee et al. 2011). Thus, microbial enzymes and even water and oxidants may be excluded or severely kinetically inhibited from entering and degrading OM or biochar itself within pores of this size. This hypothesis was tested using experimental apparatus described elsewhere (Zimmerman et al. 2004) that measures enzyme activity via the oxygen consumption that occurs when the fungal-derived enzyme laccase reacts with its substrate, 2,6-dimethoxyphenol (DMP). When the substrate was sorbed to a biochar for only a few hours, the ability of the enzyme to degrade it decreased by 79 to 96% relative to the unbound substrate, depending upon the biochar type (Fig. 4).

Another potential biochar stabilization mechanism that has been little discussed is the inactivation of degradative microbial enzymes by pyrogenic OM. There are a wide range of enzyme-mineral and enzyme-OM interactions, the majority of which, though not all, lead to enzyme inhibition (Zimmerman and Ahn 2010). We found that when laccase was sorbed to biochar's surface prior to the introduction of substrate, enzyme activity

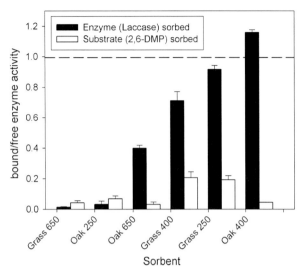

Figure 4. Enzyme activity inhibition (bound/free activity ratio) due to biochar sorption of enzyme (laccase) or sorption of substrate (2,6-dimethoxyphenol). Enzyme activity measured by O_2 consumption (see Zimmerman et al. 2004 for methods used). Biochars are described in Zimmerman (2010).

ranged from 99% less to 16% greater than the free enzyme, depending upon the biochar type (Fig. 4). Thus, even though laccase was unlikely to enter biochar's small pores, its activity was inhibited either due to conformational restriction, surface charge distortions, or pH effects, etc. These mechanisms too may explain the correspondence between BC and non-BC OM preservation as well as the formation of long-lasting fertile biochar-enriched soils.

Conclusions and Summary

Monolithic biochar degradation parameters, such as the oft quoted mean residence time of 1000–10,000 y should be abandoned in favor of an understanding of biochar as a heterogeneous material. A portion of biochar is lost with C half-lives on orders of years to decades. After this, the residual and more chemically recalcitrant component of biochar, such as that found as charcoal in ancient soils or detected as BC-soot in sediments, may have residence times of thousands to millions of years. As a practical concern, effort should be placed on developing analytical techniques that can predict the proportion of these biochar components and their associated lability, contained within a given biochar sample or naturally-derived pyrogenic material. However, much more work is needed on understanding the importance of environmental factors such as soil permeability and oxidant supply, moisture, pH and climate on biochar degradation.

Losses of biochar observed in the laboratory due to abiotic or microbial oxidation are generally in the range of 0.5 to 4% over a few years. Modeling approaches that take into account the heterogeneous nature of biochar predict losses of 3–20% over 100 y, while field studies that make use of long BC depositional records back-calculate biochar C losses in the range of 20 to 90%. While biochar leaching is a C loss mechanism that may explain this difference, it is likely that much of this C is remineralized or re-adsorbed. There is good evidence to suggest that the unaccounted for losses, which can be substantial, may be due to erosion. The particulate pyrogenic material that is exported from soil in this manor may be the source of pyrogenic C detected in natural waters.

Another area in which our understanding is sorely lacking is that of the changes to biochar that occur over time and with biochar-soil interaction. As biochar ages, through abiotic and microbial oxidation, it becomes both more soluble and, thus, even more microbially-available. With solublization, sorption and ion-bridging, it may exchange pyrogenic OM for native soil OM, while soil minerals may gain a coating of pyrogenic OM, eventually leading to physical stabilization though aggregate formation. It is these processes that surely led to the creation of *terra preta* and other soils enriched in both pyrogenic and non-pyrogenic soil C.

While this chapter has focused on biochar C losses that may occur, it must be remembered that the majority of biochar's C, which it has in abundance, will likely remain in the soil. In fact, it is likely to increase total soil C even further over time as it builds non-pyrogenic soil C both through greater plant and microbial production and through the aforementioned preservation enhancement mechanisms. Thus, we may need to revise biochar's atmosphere CO_2-C sequestration potential upward rather than downward.

References

Abiven, S., P. Hengartner, M.P.W. Schneider, N. Singh, and M.W.I. Schmidt. 2011. Pyrogenic carbon soluble fraction is larger and more aromatic in aged charcoal than in fresh charcoal. Soil Biol. Biochem. 43: 1615–1617.

Antal, M.J. and M. Gronli. 2003. The art, science, and technology of charcoal production. Industrial & Engineering Chemistry Research 42: 1619–1640.

Atkinson, C.J., J.D. Fitzgerald, and N.A. Hipps. 2010. Potential mechanisms for achieving agricultural benefits from biochar application to temperate soils: a review. Plant and Soil 337: 1–18.

Baath, E., A. Frostegard, T. Pennanen, and H. Fritze. 1995. Microbial community structure and pH response in relation to soil organic-matter quality in wood-ash fertilized, clear-cut or burned coniferous forest soils. Soil Biol. Biochem. 27: 229–240.

Baldock, J.A. and R.J. Smernik. 2002. Chemical composition and bioavailability of thermally, altered *Pinus resinosa* (Red Pine) wood. Org. Geochem. 33: 1093–1109.

Bell, M.J. and F. Worrall. 2011. Charcoal addition to soils in N.E. England: A carbon sink with environmental co-benefits? Science of the Total Environment 409: 1704–1714.

Bird, M.I. and J.A. Cali. 1998. A million-year record of fire in sub-Saharan Africa. Nature 394: 767–769.

Bird, M.I., C. Moyo, E.M. Veenendaal, J. Lloyd, and P. Frost. 1999. Stability of elemental carbon in a savanna soil. Global Biogeochem. Cycles 13: 923–932.

Brodowski, S., B. John, H. Flessa, and W. Amelung. 2006. Aggregate-occluded black carbon in soil. Eur. J. Soil Sci. 57: 539–546.

Bruun, E.W., P. Ambus, H. Egsgaard, and H. Hauggaard-Nielsen. 2012. Effects of slow and fast pyrolysis biochar on soil C and N turnover dynamics. Soil Biol. Biochem. 46: 73–79.

Bruun, S., E.S. Jensen, and L.S. Jensen. 2008. Microbial mineralization and assimilation of black carbon: Dependency on degree of thermal alteration. Org. Geochem. 39: 839–845.

Cao, X.D. and W. Harris. 2010. Properties of dairy-manure-derived biochar pertinent to its potential use in remediation. Bioresource Technology 101: 5222–5228.

Case, S.D.C., N.P. McNamara, D.S. Reay, and J. Whitaker. 2012. The effect of biochar addition on N_2O and CO_2 emissions from a sandy loam soil—The role of soil aeration. Soil Biol. Biochem. 51: 125–134.

Cheng, C.H., J. Lehmann, and M.H. Engelhard. 2008. Natural oxidation of black carbon in soils: Changes in molecular form and surface charge along a climosequence. Geochim. Cosmochem. Acta 72: 1598–1610.

Cheng, C.H., J. Lehmann, J.E. Thies, S.D. Burton, and M.H. Engelhard. 2006. Oxidation of black carbon by biotic and abiotic processes. Org. Geochem. 37: 1477–1488.

Cornelissen, G., G.D. Breedveld, S. Kalaitzidis, K. Christanis, A. Kibsgaard, and A.M.P. Oen. 2006. Strong sorption of native PAHs to pyrogenic and unburned carbonaceous geosorbents in sediments. Environ. Sci. Technol. 40: 1197–1203.

Cravo-Laureau, C., G. Hernandez-Raquet, I. Vitte, R. Jezequel, V. Belet, J.J. Godon, P. Caumette, P. Balaguer, and R. Duran. 2011. Role of environmental fluctuations and microbial diversity in degradation of hydrocarbons in contaminated sludge. Research in Microbiology 162: 888–895.

Cross, A. and S.P. Sohi. 2011. The priming potential of biochar products in relation to labile carbon contents and soil organic matter status. Soil Biol. Biochem. 43: 2127–2134.

Czimczik, C.I. and C.A. Masiello. 2007. Controls on black carbon storage in soils. Global Biogeochem. Cycles 21.

Czimczik, C.I., C.M. Preston, M.W.I. Schmidt, and E.D. Schulze. 2003. How surface fire in Siberian Scots pine forests affects soil organic carbon in the forest floor: Stocks, molecular structure, and conversion to black carbon (charcoal). Global Biogeochem. Cycles 17: 20.1–20.14.

D1762-84, A. 1990. Standard Method for Chemical Analysis of Wood Charcoal. ASTM International, Philadelphia, PA.

Dai, X., T.W. Boutton, B. Glaser, R.J. Ansley, and W. Zech. 2005. Black carbon in a temperate mixed-grass savanna. Soil Biol. Biochem. 37: 1879–1881.

Decesari, S., M.C. Facchini, E. Matta, M. Mircea, S. Fuzzi, A.R. Chughtai, and D.M. Smith. 2002. Water soluble organic compounds formed by oxidation of soot. Atmospheric Environment 36: 1827–1832.

Deenik, J.L., A. Diarra, G. Uehara, S. Campbell, Y. Sumiyoshi, and M.J. Antal. 2011. Charcoal ash and volatile matter effects on soil properties and plant growth in an acid Ultisol. Soil Sci. 176: 336–345.

Deenik, J.L., T. McClellan, G. Uehara, M.J. Antal, and S. Campbell. 2010. Charcoal volatile matter content influences plant growth and soil nitrogen transformations. Soil Sci. Soc. Am. J. 74: 1259–1270.

Ding, Y., Y.X. Liu, W.X. Wu, D.Z. Shi, M. Yang, and Z.K. Zhong. 2010. Evaluation of biochar effects on nitrogen retention and leaching in multi-layered soil columns. Water Air and Soil Pollution 213: 47–55.

Dittmar, T. 2008. The molecular level determination of black carbon in marine dissolved organic matter. Org. Geochem. 39: 396–407.

Dodson, J.R., M. Robinson, and C. Tardy. 2005. Two fine-resolution Pliocene charcoal records and their bearing on pre-human fire frequency in south-western Australia. Austral Ecol. 30: 592–599.

Focht, U. 1999. The effect of smoke from charcoal kilns on soil respiration. Environmental Monitoring and Assessment 59: 73–80.

Glaser, B. and W. Amelung. 2003. Pyrogenic carbon in native grassland soils along a climosequence in North America. Global Biogeochem. Cycles 17.

Glaser, B., E. Balashov, L. Haumaier, G. Guggenberger, and W. Zech. 2000. Black carbon in density fractions of anthropogenic soils of the Brazilian Amazon region. Org. Geochem. 31: 669–678.

Glaser, B., L. Haumaier, G. Guggenberger, and W. Zech. 2001. The 'Terra Preta' phenomenon: a model for sustainable agriculture in the humid tropics. Naturwissenschaften 88: 37–41.

Goldberg, E. 1985. Black carbon in the environment: properties and distribution. John Wiley & Sons, New York.

Gotz, G.K.E. and R.M. Fakoussa. 1999. Fungal biosolubilization of Rhenish brown coal monitored by Curie-point pyrolysis/gas chromatography/mass spectrometry using tetraethylammonium hydroxide. Appl. Microbiol. Biot. 52: 41–48.

Graber, E.R. et al. 2010. Biochar impact on development and productivity of pepper and tomato grown in fertigated soilless media. Plant and Soil 337: 481–496.

Guggenberger, G., A. Rodionov, O. Shibistova, M. Grabe, O.A. Kasansky, H. Fuchs, N. Mikheyeva, G. Zrazhevskaya, and H. Flessa. 2008. Storage and mobility of black carbon in permafrost soils of the forest tundra ecotone in Northern Siberia. Global Change Biology 14: 1367–1381.

Gustafsson, O. and P. Gschwend. 1998. The flux of black carbon to surface sediments on the New England continental shelf. Geochim. Cosmochem. Acta 62: 465–478.

Hamer, U., B. Marschner, S. Brodowski, and W. Amelung. 2004. Interactive priming of black carbon and glucose mineralisation. Org. Geochem. 35: 823–830.

Hammes, K., M.W.I. Schmidt, R.J. Smernik, L.A. Currie, W.P. Ball, T.H. Nguyen, P. Louchouarn, S. Houel, O. Gustafsson, M. Elmquist, G. Cornelissen, J.O. Skjemstad, C.A. Masiello, J. Song, P. Peng, S. Mitra, J.C. Dunn, P.G. Hatcher, W.C. Hockaday, D.M. Smith, C. Hartkopf-Froeder, A. Boehmer, B. Luer, B.J. Huebert, W. Amelung, S. Brodowski, L. Huang, W. Zhang, P.M. Gschwend, D.X. Flores-Cervantes, C. Largeau, J.N. Rouzaud, C. Rumpel, G. Guggenberger, K. Kaiser, A. Rodionov, F.J. Gonzalez-Vila, J.A. Gonzalez-Perez, J.M. de la Rosa, D.A.C. Manning, E. Lopez-Capel, and L. Ding. 2007. Comparison of quantification methods to measure fire-derived (black/elemental) carbon in soils and sediments using reference materials from soil, water, sediment and the atmosphere. Global Biogeochem. Cycles 21.

Hammes, K., M.S. Torn, A.G. Lapenas, and M.W.I. Schmidt. 2008. Centennial black carbon turnover observed in a Russian steppe soil. Biogeosciences 5: 1339–1350.

Hilscher, A., K. Heister, C. Siewert, and H. Knicker. 2009. Mineralisation and structural changes during the initial phase of microbial degradation of pyrogenic plant residues in soil. Org. Geochem. 40: 332–342.

Hilscher, A. and H. Knicker. 2011a. Carbon and nitrogen degradation on molecular scale of grass-derived pyrogenic organic material during 28 months of incubation in soil. Soil Biol. Biochem. 43: 261–270.

Hilscher, A. and H. Knicker. 2011b. Degradation of grass-derived pyrogenic organic material, transport of the residues within a soil column and distribution in soil organic matter fractions during a 28 month microcosm experiment. Org. Geochem. 42: 42–54.

Hockaday, W.C., A.M. Grannas, S. Kim, and P.G. Hatcher. 2006. Direct molecular evidence for the degradation and mobility of black carbon in soils from ultrahigh-resolution mass spectral analysis of dissolved organic matter from a fire-impacted forest soil. Org. Geochem. 37: 501–510.

Hockaday, W.C., A.M. Grannas, S. Kim, and P.G. Hatcher. 2007. The transformation and mobility of charcoal in a fire-impacted watershed. Geochim. Cosmochem. Acta 71: 3432–3445.

Hofrichter, M. et al. 1999. Degradation of lignite (low-rank coal) by ligninolytic basidiomycetes and their manganese peroxidase system. Appl. Microbiol. Biot. 52: 78–84.

Huang, P.M., T.S.C. Wang, M.K. Wang, M.H. Wu, and N.W. Hsu. 1977. Retention of phenolic acids by non-crystalline hydroxy-aluminum and hydroxy-iron compounds and clay-minerals of soils. Soil Sci. 123: 213–219.

Hyun, S. and L.S. Lee. 2004. Hydrophilic and hydrophobic sorption of organic acids by variable-charge soils: effect of chemical acidity and acidic functional group. Environ. Sci. Technol. 38: 5413–5419.

IBI. 2012. International Biochar Initiative, Standardized Product Definition and Product Testing Guidelines for Biochar That Is Used in Soil. http://www.biochar-international. org/characterizationstandard.

Inyang, M.D., B. Gao, W.C. Ding, P. Pullammanappallil, A.R. Zimmerman, and X.D. Cao. 2011. Enhanced lead sorption by biochar derived from anaerobically digested sugarcane bagasse. Separation Science and Technology 46: 1950–1956.

Jones, D.L., D.V. Murphy, M. Khalid, W. Ahmad, G. Edwards-Jones, and T.H. DeLuca. 2011. Short-term biochar-induced increase in soil CO_2 release is both biotically and abiotically mediated. Soil Biol. Biochem. 43: 1723–1731.

Jones, D.L., J. Rousk, G. Edwards-Jones, T.H. DeLuca, and D.V. Murphy. 2012. Biochar-mediated changes in soil quality and plant growth in a three year field trial. Soil Biol. Biochem. 45: 113–124.

Kaiser, K. and G. Guggenberger. 2000. The role of DOM sorption to mineral surfaces in the preservation of organic matter in soils. Org. Geochem. 31: 711–725.

Kamegawa, K., K. Nishikubo, M. Kodama, Y. Adachi, and H. Yoshida. 2002. Oxidative degradation of carbon blacks with nitric acid—II. Formation of water-soluble polynuclear aromatic compounds. Carbon 40: 1447–1455.

Karhu, K., T. Mattila, I. Bergstrom, and K. Regina. 2011. Biochar addition to agricultural soil increased CH_4 uptake and water holding capacity—Results from a short-term pilot field study. Agriculture Ecosystems & Environment 140: 309–313.

Kasozi, G.N., A.R. Zimmerman, P. Nkedi-Kizza, and B. Gao. 2010. Catechol and humic acid sorption onto a range of laboratory-produced black carbons (biochars). Environ. Sci. Technol. 44: 6189–6195.

Keil, R.G., D.B. Montluçon, F.G. Prahl, and J.I. Hedges. 1994. Sorptive preservation of labile organic matter in marine sediments. Nature 370: 549–551.

Keiluweit, M., P.S. Nico, M.G. Johnson, and M. Kleber. 2010. Dynamic molecular structure of plant biomass-derived black carbon (biochar). Environ. Sci. Technol. 44: 1247–1253.

Keith, A., B. Singh, and B.P. Singh. 2011. Interactive Priming of Biochar and Labile Organic Matter Mineralization in a Smectite-Rich Soil. Environ. Sci. Technol. 45: 9611–9618.

Kercher, A.K. and D.C. Nagle. 2003. Microstructural evolution during charcoal carbonization by X-ray diffraction analysis. Carbon 41: 15–27.

Khodadad, C.L.M., A.R. Zimmerman, S.J. Green, S. Uthandi, and J.S. Foster. 2011. Taxa-specific changes in soil microbial community composition induced by pyrogenic carbon amendments. Soil Biol. Biochem. 43: 385–392.

Kim, S.W., L.A. Kaplan, R. Benner, and P.G. Hatcher. 2004. Hydrogen-deficient molecules in natural riverine water samples—evidence for the existence of black carbon in DOM. Mar. Chem. 92: 225–234.

Kleber, M. 2010. What is recalcitrant soil organic matter? Environ. Chem. 7: 320–332.

Knicker, H. 2007. How does fire affect the nature and stability of soil organic nitrogen and carbon? A review. Biogeochemistry 85: 91–118.

Knicker, H. 2011. Pyrogenic organic matter in soil: Its origin and occurrence, its chemistry and survival in soil environments. Quaternary International 243: 251–263.

Knicker, H., G. Almendros, F.J. Gonzalez-Vila, J.A. Gonzalez-Perez, and O. Polvillo. 2006. Characteristic alterations of quantity and quality of soil organic matter caused by forest fires in continental Mediterranean ecosystems: a solid-state (13)C NMR study. Eur. J. Soil Sci. 57: 558–569.

Knicker, H., A. Hilscher, F.J. Gonzalez-Vila, and G. Almendros. 2008. A new conceptual model for the structural properties of char produced during vegetation fires. Org. Geochem. 39: 935–939.

Knicker, H., P. Muller, and A. Hilscher. 2007. How useful is chemical oxidation with dichromate for the determination of "Black Carbon" in fire-affected soils? Geoderma 142: 178–196.

Knicker, H., K.U. Totsche, G. Almendros, and F.J. Gonzalez-Vila. 2005. Condensation degree of burnt peat and plant residues and the reliability of solid-state VACP MAS C-13 NMR spectra obtained from pyrogenic humic material. Org. Geochem. 36: 1359–1377.

Kolb, S.E., K.J. Fermanich, and M.E. Dornbush. 2009. Effect of charcoal quantity on microbial biomass and activity in temperate soils. Soil Sci. Soc. Am. J. 73: 1173–1181.

Kuhlbusch, T.A. and P.J. Crutzen. 1995. Toward a global estimate of black carbon in residues of vegetation fires representing a sink of atmospheric CO_2 and a source of O_2. Global Biogeochem. Cycles 9: 491–501.

Kuhlbusch, T.A.J. 1995. Method for determining black carbon in residues of vegetation fires. Environ. Sci. Technol. 29: 2695–2702.

Kurth, V.J., M.D. MacKenzie, and T.H. DeLuca. 2006. Estimating charcoal content in forest mineral soils. Geoderma 137: 135–139.

Kuzyakov, Y., J.K. Friedel, and K. Stahr. 2000. Review of mechanisms and quantification of priming effects. Soil Biol. Biochem. 32: 1485–1498.

Kuzyakov, Y., I. Subbotina, H.Q. Chen, I. Bogomolova, and X.L. Xu. 2009. Black carbon decomposition and incorporation into soil microbial biomass estimated by C-14 labeling. Soil Biol. Biochem. 41: 210–219.

Lehmann, J., C. Czimczik, D. Laird, and S. Sohi. Stability of biochar in the soil. in: Lehmann, J.,Joseph, S. (Eds.). 2009. Biochar for Environmental Management: Science and Technology. Earthscan, London, pp. 183–205.

Lehmann, J. et al. 2011. Biochar effects on soil biota: A review. Soil Biol. Biochem.

Lehmann, J. et al. 2008. Spatial complexity of soil organic matter forms at nanometre scales. Nature Geoscience 1: 238–242.

Lepik, R. and T. Tenno. 2012. Determination of biodegradability of phenolic compounds, characteristic to wastewater of the oil-shale chemical industry, on activated sludge by oxygen uptake measurement. Environ. Technol. 33: 329–339.

Liang, B. et al. 2008. Stability of biomass-derived black carbon in soils. Geochim. Cosmochem. Acta 72: 6069–6078.

Liang, B.Q., J. Lehmann, S.P. Sohi, J.E. Thies, B. O'Neill, L. Trujillo, J. Gaunt, D. Solomon, J. Grossman, E.G. Neves, and F.J. Luizao. 2010. Black carbon affects the cycling of non-black carbon in soil. Org. Geochem. 41: 206–213.

Luo, Y., M. Durenkamp, M. De Nobili, Q. Lin, and P.C. Brookes. 2011. Short term soil priming effects and the mineralisation of biochar following its incorporation to soils of different pH. Soil Biol. Biochem. 43: 2304–2314.

Major, J., J. Lehmann, M. Rondon, and C. Goodale. 2010. Fate of soil-applied black carbon: downward migration, leaching and soil respiration. Global Change Biology 16: 1366–1379.

Mannino, A. and H.R. Harvey. 2004. Black carbon in estuarine and coastal ocean dissolved organic matter. Limnology and Oceanography 49: 735–740.

Masiello, C.A. 2004. New directions in black carbon organic geochemistry. Mar. Chem. 92: 201–213.

Masiello, C.A. and E.R.M. Druffel. 2003. Organic and black carbon C-13 and C-14 through the Santa Monica Basin sediment oxic-anoxic transition. Geophys. Res. Lett. 30: 1185.

Mayer, L.M., L.L. Schick, K.R. Hardy, R. Wagal, and J. McCarthy. 2004. Organic matter in small mesopores in sediments and soils. Geochim. Cosmochem. Acta 68: 3863–3872.

Middleburg, J.J. 1989. A simple rate model for organic matter decomposition in marine sediments. Geochim. Cosmochem. Acta 53: 1577–1581.

Middleburg, J.J., J. Nieuwenhuize, and P. van Breugel. 1999. Black carbon in marine sediments. Mar. Chem. 65: 245–252.

Mitra, S., T.S. Bianchi, B.A. McKee, and M. Sutula. 2002. Black carbon from the Mississippi River: Quanities, sources, and potential implications for the global carbon cycle. Environ. Sci. Technol. 36: 2296–2302.

Mitra, S., A.R. Zimmerman, G.B. Hunsinger, D. Willard, and J.C. Dunn. 2009. A Holocene record of climate-driven shifts in coastal carbon sequestration. Geophys. Res. Lett. 36.

Mukherjee, A. and A.R. Zimmerman. In press. Organic carbon and nutrient release from a range of laboratory-produced biochars and biochar–soil mixtures. Geoderma.

Mukherjee, A. 2011. PhD Dissertation: Physical and chemical properties of a range of laboratory-produced fresh and aged biochars, University of Florida, Gainesville, FL, 127 pp.

Mukherjee, A., A.R. Zimmerman, and W. Harris. 2011. Surface chemistry variations among a series of laboratory-produced biochars. Geoderma 163: 247–255.

Murage, E.W., P. Voroney, and R.P. Beyaert. 2007. Turnover of carbon in the free light fraction with and without charcoal as determined using the C-13 natural abundance method. Geoderma 138: 133–143.

Nguyen, B.T. and J. Lehmann. 2009. Black carbon decomposition under varying water regimes. Org. Geochem. 40: 846–853.

Nguyen, B.T., J. Lehmann, W.C. Hockaday, S. Joseph, and C.A. Masiello. 2010. Temperature sensitivity of black carbon decomposition and oxidation. Environ. Sci. Technol. 44: 3324–3331.

Nguyen, B.T. et al. 2008. Long-term black carbon dynamics in cultivated soil. Biogeochemistry 89: 295–308.

Novak, J.M. et al. 2010. Short-term CO_2 mineralization after additions of biochar and switchgrass to a Typic Kandiudult. Geoderma 154: 281–288.

Ohlson, M., B. Dahlberg, T. Okland, K.J. Brown, and R. Halvorsen. 2009. The charcoal carbon pool in boreal forest soils. Nature Geoscience 2: 692–695.

O'Neill, B. et al. 2009. Bacterial community composition in Brazilian Anthrosols and adjacent soils characterized using culturing and molecular identification. Microbial Ecology 58: 23–35.

Pederson, D.C., D.M. Peteet, D. Kurdyla, and T. Guilderson. 2005. Medieval Warming, Little Ice Age, and European impact on the environment during the last millennium in the lower Hudson Valley, New York, USA. Quat. Res. 63: 238–249.

Pessenda, L.C.R., S.E.M. Gouveia, and R. Aravena. 2001. Radiocarbon dating of total soil organic matter and humin fraction and its comparison with C-14 ages of fossil charcoal. Radiocarbon 43: 595–601.

Pessenda, L.C.R. et al. 2005. Holocene palaeoenvironmental reconstruction in northeastern Brazil inferred from pollen, charcoal and carbon isotope records. Holocene 15: 812–820.

Potter, M.C. 1908. Bacteria as agents in the oxidation of amorphous carbon. Proc. R. Soc. London, Ser. B. 80: 239–259.

Pradhan, B., S. Murugavelh, and K. Mohanty. 2012. Phenol biodegradation by indigenous mixed microbial consortium: Growth kinetics and inhibition. Environmental Engineering Science 29: 86–92.

Reiss, J. 1992. Studies on the Solubilization of German Coal by Fungi. Appl. Microbiol. Biot. 37: 830–832.

Rodriguez-Navarro, C., E. Ruiz-Agudo, A. Luque, A.B. Rodriguez-Navarro, and M. Ortega-Huertas. 2009. Thermal decomposition of calcite: Mechanisms of formation and textural evolution of CaO nanocrystals. American Mineralogist 94: 578–593.

Rumpel, C. et al. 2006. Black carbon contribution to soil organic matter composition in tropical sloping land under slash and burn agriculture. Geoderma 130: 35–46.

Rumpel, C., A. Ba, F. Darboux, V. Chaplot, and O. Planchon. 2009. Erosion budget and process selectivity of black carbon at meter scale. Geoderma 154: 131–137.

Santos, F., M.S. Torn, and J.A. Bird. 2012. Biological degradation of pyrogenic organic matter in temperate forest soils. Soil Biol. Biochem. 51: 115–124.

Saravanan, P., K. Pakshirajan, and P.K. Saha. 2008. Kinetics of growth and multi substrate degradation by an indigenous mixed microbial culture isolated from a wastewater treatment plant in Guwahati, India. Water Air Soil Pollut. 58: 1101–1106.

Schmidt, M.W.I., J.O. Skjemstad, and C. Jager. 2002. Carbon isotope geochemistry and nanomorphology of soil black carbon: Black chernozemic soils in central Europe originate from ancient biomass burning. Global Biogeochem. Cycles 16.

Schneider, M.P.W., M. Hilf, U.F. Vogt, and M.W.I. Schmidt. 2010. The benzene polycarboxylic acid (BPCA) pattern of wood pyrolyzed between 2000°C and 1000°C. Org. Geochem. 41: 1082–1088.

Schneider, M.P.W., J. Lehmann, and M.W.I. Schmidt. 2011. Charcoal quality does not change over a century in a tropical agro-ecosystem. Soil Biol. Biochem. 43: 1992–1994.

Scott, A.C. 2000. The Pre-Quaternary history of fire. Palaeogeography Palaeoclimatology Palaeoecology 164: 281–329.

Scott, C.D., G.W. Strandberg, and S.N. Lewis. 1986. Microbial solubilization of coal. Biotechnol. Progr. 2: 131–139.

Sharma, R.K. et al. 2004. Characterization of chars from pyrolysis of lignin. Fuel 83: 1469–1482.

Shneour, E.A. 1966. Oxidation of graphitic carbon in certain soils. Science 151: 991.

Singh, B., B.P. Singh, and A.L. Cowie. 2010a. Characterisation and evaluation of biochars for their application as a soil amendment. Aust. J. Soil Res. 48: 516–525.

Singh, B.P., B.J. Hatton, B. Singh, A.L. Cowie, and A. Kathuria. 2010b. Influence of biochars on nitrous oxide emission and nitrogen leaching from two contrasting soils. J. Env. Qual. 39: 1224–1235.

Skjemstad, J.O., D.C. Reicosky, A.R. Wilts, and J.A. McGowan. 2002. Charcoal carbon in US agricultural soils. Soil Sci. Soc. Am. J. 66: 1249–1255.

Skjemstad, J.O., J.A. Taylor, and R.J. Smernik. 1999. Estimation of charcoal (Char) in soils. Communications In Soil Science and Plant Analysis 30: 2283–2298.

Sollins, P., P. Homann, and B.A. Caldwell. 1996. Stabilization and destabilization of soils organic matter: mechansims and controls. Geoderma 74: 65–105.

Song, J., P.A. Peng, and W. Huang. 2002. Black carbon and kerogen in soils and sediments. 1. Quantification and characterization. Environ. Sci. Technol. 36: 3960–3967.

Spokas, K.A. 2010. Review of the stability of biochar in soils: predictability of O:C molar ratios. Carbon Management 1.

Spokas, K.A., W.C. Koskinen, J.M. Baker, and D.C. Reicosky. 2009. Impacts of woodchip biochar additions on greenhouse gas production and sorption/degradation of two herbicides in a Minnesota soil. Chemosphere 77: 574–581.

Spokas, K.A. et al. 2011. Qualitative analysis of volatile organic compounds on biochar. Chemosphere 85: 869–882.

Spokas, K.A. and D.C. Reicosky. 2009. Impacts of sixteen different biochars on soil greenhouse gas production. Plant Soil 3: 179–193.

Steinbeiss, S., G. Gleixner, and M. Antonietti. 2009. Effect of biochar amendment on soil carbon balance and soil microbial activity. Soil Biol. Biochem. 41: 1301–1310.

Steiner, C., K.C. Das, M. Garcia, B. Forster, and W. Zech. 2008. Charcoal and smoke extract stimulate the soil microbial community in a highly weathered xanthic Ferralsol. Pedobiologia 51: 359–366.

Sun, M.Y., R.C. Aller, C. Lee, and S.G. Wakeham. 2002. Effects of oxygen and redox oscillation on degradation of cell-associated lipids in surficial marine sediments. Geochim. Cosmochem. Acta 66: 2003–2012.

Thies, J.E. and M.C. Rillig. Characterisitics of biochar: Bioogical properties. *In:* J. Lehmann, S. Joseph (Eds.). 2009. Biochar for Environmental Management: Science and Technology. Earthscan, London, pp. 183–205.

Vasilyeva, N.A. et al. 2011. Pyrogenic carbon quantity and quality unchanged after 55 years of organic matter depletion in a Chernozem. Soil Biol. Biochem. 43: 1985–1988.

Wang, Y. et al. 2011. *In situ* degradation of phenol and promotion of plant growth in contaminated environments by a single Pseudomonas aeruginosa strain. Journal of Hazardous Materials 192: 354–360.

Wardle, D.A., M.C. Nilsson, and O. Zackrisson. 2008. Fire-derived charcoal causes loss of forest humus. Science 320: 629–629.

Warnock, D.D., J. Lehmann, T.W. Kuyper, and M.C. Rillig. 2007. Mycorrhizal responses to biochar in soil—concepts and mechanisms. Plant and Soil 300: 9–20.

Wengel, M., E. Kothe, C.M. Schmidt, K. Heide, and G. Gleixner. 2006. Degradation of organic matter from black shales and charcoal by the wood-rotting fungus Schizophyllum commune and release of DOC and heavy metals in the aqueous phase. Science of the Total Environment 367: 383–393.

Woolf, D., J.E. Amonette, F.A. Street-Perrott, J. Lehmann, and S. Joseph. 2010. Sustainable biochar to mitigate global climate change 1: 56.

Yang, L. et al. 2011. Promotion of plant growth and in situ degradation of phenol by an engineered Pseudomonas fluorescens strain in different contaminated environments. Soil Biol. Biochem. 43: 915–922.

Yuan, J.H., R.K. Xu, and H. Zhang. 2011. The forms of alkalis in the biochar produced from crop residues at different temperatures. Bioresource Technology 102: 3488–3497.

Zimmerman, A. 2010. Abiotic and microbial oxidation of laboratory-produced black carbon (biochar). Environ. Sci. Technol. 44: 1295–1301.

Zimmerman, A.R., B. Gao, and M.-Y. Ahn. 2011. Positive and negative carbon mineralization priming effects among a variety of biochar-amended soils. Soil Biol. Biochem. 43: 1169–1179.

Zimmerman, A.R. and M.-Y. Ahn. Organo-mineral enzyme interactions and influence on soil enzyme activity. *In:* G.C. Shukla, A. Varma (Eds.). 2010. Soil Enzymes. Springer, Berlin Heidelberg.

Zimmerman, A.R., J. Chorover, K.W. Goyne, and S.L. Brantley. 2004. Protection of mesopore-adsorbed organic matter from enzymatic degradation. Environ. Sci. Technol. 38: 4542–4548.

Biochar Impact on Plant Resistance to Disease

E.R. Graber[1,*] and Y. Elad[2]

1 **Why Biochar?**

2 **Plant Disease**

   2.1 **General**

   2.2 **Foliar Diseases**

   2.3 **Soil-borne Diseases**

   2.4 **Systemic Plant Responses to Disease**

3 **Biochar-mediated Protection of Plants from Foliar and Soil-borne Disease**

   3.1 **Possible Mechanisms by which Biochar May Protect Plants against Diseases**

      3.1.1 **Improved Nutrient Supply and Enhanced Plant Growth**

      3.1.2 **Increase in Beneficial Soil Microorganisms**

      3.1.3 **Sorption of Toxins Produced by Disease Pathogens**

      3.1.4 **Suppression of Soil-borne Pathogens by Biochar-Derived Toxic Organic Compounds**

[1]Department of Soil Chemistry, Plant Nutrition and Microbiology, Institute of Soil, Water and Environmental Sciences, The Volcani Center, Agricultural Research Organization, Bet Dagan 50250 Israel; E-mail: ergraber@volcani.agri.gov.il

[2]Department of Plant Pathology and Weed Research, Institute of Plant Protection, The Volcani Center, Agricultural Research Organization, Bet Dagan 50250 Israel; E-mail: elady@volcani.agri.gov.il

*Corresponding author

List of glossary given at the end of the text.

1 Why Biochar?

Since 2007, there has been the remarkable development worldwide of an agenda promoting the use of charcoal as a soil amendment. This agenda is based on four main pillars: (i) *Renewable Energy Production*: Charcoal (a solid energy product) is produced by thermally degrading biomass at temperatures above ~300°C in the absence of oxygen, a process known as pyrolysis. During biomass pyrolysis, gaseous (syngas consisting of H_2, CH_4, CO, CO_2, and C_2H_4 and other small hydrocarbons) and liquid (bio-oil) energy products are also produced. The pyrolysis technology is poised to become part of an arsenal of technologies, including production of biodiesel from oil-rich crops, ethanol from cellulosic crops, and biogas from anaerobic digestion of wastes, for producing renewable energy from biomass. The goal is to replace some fossil fuels with these different biofuels in order to lower net greenhouse gas emissions and to diversify energy supplies. (ii) *Waste Treatment*: Many various wastes, including agricultural, forestry, food industry, and urban biomass wastes, can be treated by pyrolysis and thus converted into energy products. As a result, pyrolysis is considerably more versatile than biofuel technologies which require purpose-grown crops that compete with food production for resources. (iii) *Soil Fertility*: Charcoal used as a soil additive together with organic and inorganic fertilizers has the ability to significantly improve soil fertility by improving soil physical and chemical attributes. Therefore, soil amendment by charcoal may help ward off soil degradation and restore already degraded soils, assisting in the establishment of sustainable food and fuel production in areas with severely depleted soils and scarce resources. (iv) *Carbon Sequestration*: Depending on feedstock and pyrolysis conditions, biochar is estimated to have a half-life in soil of hundreds to tens of thousands of years, leading to long-term, below-ground sequestration of carbon that originated as CO_2 in the atmosphere. Moreover, addition of biochar to soil may reduce greenhouse gas emissions from cultivated soils. These four pillars together constitute the "Charcoal Vision" (Laird 2008). To differentiate this 4-fold paradigm from the conventional view of charcoal solely as an energy product, a new name for it has emerged: BIOCHAR.

The pyrolysis/biochar platform is in its infancy, and biochar has yet to enjoy widespread use in modern agriculture. There are still a number of impediments to the adoption of biochar as a soil amendment, including technical issues, cost, and a lack of understanding of how biochar physical and chemical characteristics affect its utility in improving soil fertility. Biochar characteristics can vary considerably depending on the initial feedstock and production parameters (Keiluweit et al. 2010). In particular, the temperature at which biochar is prepared has a profound impact on its characteristics, with biochar produced at relatively lower temperatures having lower pH values, lower Specific Surface Areas (SSA), and higher Cation Exchange Capacities (CEC) per unit surface area than those prepared at relatively higher temperatures (Gaskin et al. 2008). The ways in which these characteristics influence biochar suitability as a soil amendment are still unknown. These qualities can also affect biochar stability in the soil, thus having an impact on its value as a long term carbon sink.

Despite being little applied in modern agriculture, biochar was an important additive used to improve soil fertility over several millennia. The pre-Columbian natives of the Amazon Basin added charcoal together with manure, bones and pottery shards to their infertile soils, creating productive anthrosols which have retained their fertility to this day (Sombroek 1966). The nutrient-holding capacity of the added charcoal is considered the major factor responsible for the persistent fertility of these "Terra Preta" soils, though abandoned 100s to 1000s of years ago (Smith 1980). Similar pockets of ancient, fertile, charcoal-containing anthrosols scattered amongst soils of low native fertility have also been found in other locations around the world, including Ecuador, Peru, Western Africa, South Africa, Australia, and Asia. Moreover, charcoal enjoyed widespread use in 19th and early 20th century agriculture and horticulture in North America and Europe for a variety of reasons, including its ability to retain nutrients, hasten seed germination, and improve seedling vigor (Allen 1847). Charcoal was also known to help crops withstand diseases caused by different foliar fungal pathogens:

> *"Charcoal […] often checks rust in wheat, and mildew in other crops; and in all cases mitigates their ravages, where it does not wholly prevent them (p. 45).*
>
> *A dressing of charcoal has in many instances, been found an adequate preventative [of rust]; and so beneficial has it proved in France, that it has been extensively introduced there for the wheat crop (p. 109)."*
> *(Allen 1847).*

The ability of biochar added to soil to promote plant health and improve plant resistance to plant disease-causing pathogens is explored in this chapter.

2 Plant Disease

2.1 General

Pathogens and pests cause substantial losses in the amount and quality of crop yield. Plant pathogens include species of fungi, chromalveolata (fungus-like organisms), bacteria, viruses, viroids and phytoplasma. These pathogens fall into two broad groups: obligate parasites, which depend entirely on a living host plant tissue for their nutrition and reproduction, and facultative parasites, which can cause considerable damage to plants, but can also live as saprophytes on plant residues and organic material. Obligate fungal parasites include many pathogens that cause diseases such as powdery mildews, downy mildews and rusts. The names of these disease groups generally describe the disease symptoms. Facultative parasites include soil-borne pathogens such as *Pythium* spp., *Rhizoctonia solani*, *Sclerotium* spp. and *Fusarium* spp. that cause damping off, root rot, stem-base rot, and plant wilt, as well as foliar pathogens such as the ubiquitous gray mold fungus *Botrytis cinerea*, and *Alternaria* spp., that causes leaf spots. There are also some semi-biotrophic pathogens that colonize tissue and depend on living host plants, but also have some ability to utilize organic material. Important semi-biotrophic pathogens include *Phytophthora infestans* that causes late blight in tomato and potato, and the anthracnose fungi *Colletotrichum* spp.

Pathogens are common in air (mainly as conidia), soil, and on plant surfaces throughout the world in arid, tropical, temperate and alpine regions, and most plants are susceptible to some pathogens. Pathogens can be divided into two groups: soil-borne and foliar, according to the plant organs they infect and the habitat in which they thrive. Fungi and bacteria penetrate host plant tissues through natural openings such as the stomata; viruses, fungi, and bacteria also enter through wounds. Moreover, many fungi can penetrate leaves or roots directly without using wounds or stomata.

All fungi and fungus-like pathogens form microscopic conidia or some sort of condensed, hard, hyphal structures that serve as pathogen dispersal agents and/or survival structures. Generally, wet or humid weather and wet soil are important in the development of many fungal and bacterial diseases. After landing on plant foliar surfaces or close to its roots, the conidia and other structures germinate and penetrate the susceptible host tissue and initiate the disease. Changing the biotic and abiotic plant environment as well as the susceptibility of plant organs to pathogen infection are some of the useful means of disease suppression that are available and can substitute the use of chemical pesticides.

2.2 Foliar Diseases

Diseases that are caused by pathogens which infect above-ground plant organs, i.e., the stem, leaves, flowers and fruits, are termed foliar diseases. The symptoms caused by foliar pathogens include rot and mold, spots, malformation, chlorosis and discoloration on the various upper plant organs. Some foliar pathogens like *B. cinerea*, *Colletotrichum* spp., *Monilinia* spp., and *Alternaria* spp. are widely distributed in cultivated areas and cause damage to many plant species, whereas other pathogens like *P. infestans* and those grouped among the powdery mildews, rusts, and downy mildews, though very widespread, are specific to host species.

Foliar diseases may be divided into those promoted by high humidity and those that develop under conditions of lower humidity. High humidity, free water film on susceptible organs, reduced intensity of light, and moderate temperatures promote the development of plant diseases caused by pathogens of the first group, since their conidia are better able to germinate, penetrate and establish infection under these conditions. These conditions are often amplified by the development of a luxuriant plant canopy, which reduces aeration and illumination and facilitates the development of diseases. The common humidity-promoted fungal diseases that thrive under these conditions are gray mold (*B. cinerea*); tomato and potato late blight (*P. infestans*), downy mildews, tomato leaf mold (*Fulvia fulva*), white mold (*Sclerotinia sclerotiorum*) and leaf spots (*Alternaria*, *Septoria* and *Cercospora* spp.). Germination of conidia of pathogens of the second group is promoted by higher humidity but they can also germinate and infect at medium or even low humidity and their symptoms further develop at lower humidity conditions. Pathogens included in the second group are those which cause powdery mildews (genera like *Sphaerotheca*, *Podosphaera*, *Leveillula*); they are favored by high humidity, but leaf surface wetness is detrimental to their conidia viability.

2.3 Soil-borne Diseases

Diseases that are caused by pathogens which survive in the soil and in plant residues are regarded as soil-borne diseases. Many of these pathogens are widely distributed in cultivated areas and cause damage to root, stem and crown tissues. The symptoms are usually apparent when the above-ground parts of the plant die or show symptoms such as stunting, wilting and yellowing, caused because of disruption in the uptake and translocation of water and nutrients from the soil. Soil-borne pathogens can survive for a long time in the absence of a susceptible host. Some soil-borne species persist as resistant hyphae in plant residues or as propagules in form of thick walled spores such as oospores and chlamydospores, sclerotia and microsclerotia; all are adapted for long-term survival in soil, and common control measures are not always effective for their suppression. Some soil pathogens have a wide host range whereas others have a very narrow host range.

The various types of disease symptoms can be grouped into categories. Seedling blight and damping-off are commonly caused by *Pythium, Phytophthora, Rhizoctonia* and *Sclerotium rolfsii.* These pathogens can infect seedlings during germination, pre-emergence or post-emergence phases of seedling establishment. Genera such as *Pythium, Phytophthora, Rhizoctonia, Cylindrocladium, Armillaria, Dematophora,* and *Fusarium* cause root rot that is characterized by decay of the true root system either by attacking the juvenile roots or the older parts of the root system. Stem and collar rots are caused by species belonging to genera such as *Phytophthora, Sclerotium, Rhizoctonia, Sclerotinia,* and *Fusarium,* causing decay of the stem at ground level that may lead to symptoms of wilting, death of leaves, and death of the plant. *Fusarium oxysporum* and *Verticillium* spp. cause wilt diseases.

2.4 Systemic Plant Responses to Disease

Under certain conditions, plants can exhibit innate, system-wide resistance against disease pathogens. These defenses, known as Induced Resistance (IR), can be effective against a wide range of pathogens and parasites including fungi, bacteria, viruses and nematodes. IR is defined as a physiological state of enhanced defensive capacity which typically is elicited by specific stimuli, and whereupon the plant's innate defenses are potentiated against subsequent challenges (Vallad and Goodman 2004). In model plant systems, induced resistance is conveniently subdivided into two major types: Systemic Acquired Resistance (SAR) and Induced Systemic Resistance (ISR). SAR and ISR are commonly differentiated by their regulatory pathways and the nature of the elicitors.

SAR, effective for a variety of plant species, is associated with the production of Pathogenesis-Related (PR) proteins and mediated via a Salicylic Acid (SA)-dependent process. It can be triggered by both chemical elicitors and biological elicitors (virulent, avirulent, and nonpathogenic microorganisms) (Vallad and Goodman 2004). Chemical elicitors include the synthetic salicylic acid analogues 2,6-dichloroisonicotinic acid and acibenzolar-S-methyl (Iriti et al. 2004, Perazzolli et al. 2008), methyl jasmonate (Belhadj et al. 2006), chitin (Rajkumar et al. 2008b) and chitosan (Aziz et al. 2006), laminarin (Trouvelot et al. 2008), and β-aminobutyric acid (Hamiduzzaman et al. 2005). In addition, systemic resistance can be stimulated by phosphate salts, silicon, amino acids, fatty acids, and cell wall fragments (Reuveni et al. 1995, Walters et al. 2005, Wiese et al. 2005). Biological elicitors can be chemicals released by micro-organisms that induce SAR along the SA pathway, such as seen for members of the *Trichoderma* genera (Harman et al. 2004). Environmental agents such as osmotic, moisture and proton stresses, mechanical wounding, and temperature extremes can also induce SAR (Ayres 1984, Wiese et al. 2004).

In contrast, ISR develops systemically in response to colonization of plant roots by Plant Growth-Promoting Rhizobacteria (PGPR) and Fungi (PGPF) (Van der Ent et al. 2009), and is mediated by the phytohormones Jasmonic Acid (JA) and Ethylene (E). The ability of PGPR and PGPF to promote ISR is specific to certain plant species and genotypes (Van Wees et al. 1997). ISR does not involve expression of PR proteins (Van Loon et al. 1998).

Disease resistance is often associated with an overall heightened capacity of the plant to induce cellular defenses when a stress is first encountered, and is known as "the primed state of the plant" (Ton and Maunch-Mani 2003). Primed plants display faster and stronger activation of cellular defense responses following attack by a pathogen as compared with non-primed plants (Conrath et al. 2006). These heightened responses can involve both earlier oxidative burst and stronger up-regulation of defense genes (Conrath et al. 2002, Ahn et al. 2007). The physiological and molecular mechanisms that underlie primed responses are not well understood, but priming has been observed to be an integral part of both SAR and ISR. Moreover, the priming effect is not restricted to biotic stresses, but has been observed also for abiotic stresses such as salt, heat, cold and drought (Ton and Maunch-Mani 2003).

3 Biochar-mediated Protection of Plants from Foliar and Soil-borne Disease

The first modern evidence that biochar added to the soil medium was capable of mediating plant systemic resistance against foliar disease-causing microorganisms was recorded in greenhouse pepper and tomato plants (Elad et al. 2010). The severity of diseases caused by necrotrophic (*B. cinerea*) and biotrophic (*Oidiopsis sicula*, originally referred to according to its teleomorph name, *Leveillula taurica*) foliar pathogens was significantly reduced in biochar-amended treatments (Fig. 1).

Reduced damage by broad mite (*Polyphagotarsonemus latus*) in biochar-amended pepper plants was also observed (Elad et al. 2010). Suppression of *Podospaera aphanis* (biotrophic), *B. cinerea* (necrotrophic) and *Colletotrichum acutatum* (semi-biotrophic) on the leaves of strawberry plants grown in

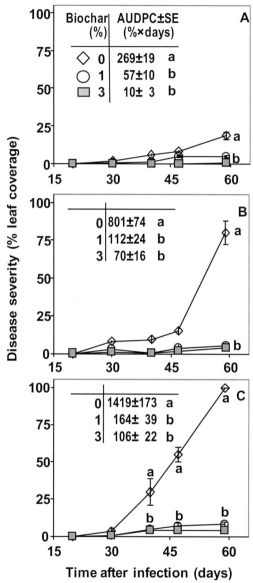

Figure 1. Effect of biochar mixed in potting medium on biotrophic foliar pathogen (*Oidiopsis sicula*, originally referred to according to its teleomorph name, *Leveillula taurica*) that causes powdery mildew in tomato plants inoculated 60 days after planting and evaluated on high (A), medium (B) and lower (C) leaves. Disease is presented as % of leaf coverage, and as area under disease pressure curves (AUDPC±SE) through 59 days. Plants were incubated under conditions of 20–30°C. Bars represent the standard error of the mean of six replicates. Data points labeled by a common letter are not significantly different according to Fisher's protected LSD test. Reprinted with permission from Elad et al. (2010).

biochar-amended soils has also been reported (Meller Harel et al. 2012). Considering that these disease-causing fungi have very different infection strategies, it is apparent that biochar can exhibit a broad capacity for controlling foliar fungal diseases. Being that the site of the biochar (in the soil) during all stages of plant development was spatially separate from the site of infection (the leaves), and that at no point did the life-cycle of the disease microorganisms pass through the soil, it is evident that (i) there was no direct toxicity of the biochar towards the causal agents, and (ii) the presence of biochar in the soil medium caused a system-wide response in the plants. Two different biochars, one from wood and the other from greenhouse waste showed similar effects. Plant nutrition and water balance were not factors in the reduced susceptibility to the diseases (Graber et al. 2010).

Using real-time qPCR techniques, molecular evidence for the induction by biochar of systemic strawberry plant defenses against the foliar pathogens *B. cinerea* and *P. aphanis* was discovered (Meller Harel et al. 2011, Meller Harel et al. 2012). By following the relative expression of five defense-related genes (FaPR1, Faolp2, Fra a3, Falox, and FaWRKY1) in strawberry leaves, it was shown that biochar addition to the potting medium triggered both salicylic acid-induced (SAR) and jasmonic acid/ethylene-induced (ISR) gene expression in the leaves. Moreover, plants grown in the biochar amended mix were primed for gene expression upon infection by *B. cinerea* and by *P. aphanis*. The ability of biochar to mediate gene expression along both SAR and ISR pathways may explain why it helped reduce infection by pathogens having different infection strategies.

In so much as biochars have highly variable physical and chemical properties depending on original feedstock and pyrolysis conditions (Amonette and Joseph 2009, Downie et al. 2009, Krull et al. 2009), it still needs to be determined which types of biochar induce systemic resistance responses. The effect of slow pyrolysis biochars produced at two Highest Treatment Temperatures (HTT; 350 and 450°C) from three biomass feedstocks (greenhouse waste, olive pomace, and eucalyptus wood) on infection of tomato leaves by *B. cinerea* was reported (Elad et al. 2011). In most cases, the biochars induced resistance towards gray mold regardless of the feedstock, HTT, type of disease assay performed (whole plant attached leaves vs. detached leaves), period of exposure to the biochar, and plant age. Nevertheless, it can be expected that disease control efficacy will vary with other biochar production temperatures, biomass sources, plant growth systems, plant species and diseases.

Besides inducing systemic plant resistance responses to foliar fungal pathogens, evidence for biochar-induced suppressiveness of soil disease via its influence on arbuscular mycorrhizal fungi (AMF) was presented by Matsubara et al. (2002). They demonstrated that biochar amendments had a suppressive effect on crown and root rot caused by the soil-borne pathogen *Fusarium*, and increased AM colonization of asparagus seedlings. Biochar added to asparagus field soil was also found to result in reductions in root lesions caused by *Fusarium oxysporum* f. sp. *asparagi* and *F. proliferatum* (Elmer and Pignatello 2011). In our recent unpublished work, eucalyptus wood biochar was found to suppress root rot caused by *R. solani* in tomato and cucumber (Frenkel et al. unpublished results).

3.1 Possible Mechanisms by which Biochar May Protect Plants against Diseases

Given the complexity of the soil-rhizosphere-plant continuum, the intimate inter-relationships between soil chemistry, biology and structure, and the immature status of most avenues of biochar research, there is more unknown than known in terms of the mechanisms by which biochar can induce systemic plant defenses or enhance soil suppressiveness. The following list is an admittedly incomplete enumeration of some of the possible mechanisms:

i) Biochar may provide nutrients and improve nutrient solubilization and uptake, thus enhancing plant growth and resistance against pathogenic microorganisms.

ii) Biochar addition frequently results in an increase in soil microbial biomass and causes changes in the soil microbial community. These changes can result in direct protection against soil pathogens via microbial production of antibiotics (antibiosis), competition for resources, or parasitism of the disease microorganism. Beneficial microorganisms can also directly enhance plant growth.

iii) Biochar may absorb the toxins produced by soil disease microorganisms, such as extracellular enzymes and organic acids, thus protecting the plant from physical ravages by soil-borne pathogens.

iv) Toxic organic compounds associated with the labile fraction of biochar may suppress disease microorganisms in the soil.

v) Biochar may change the chemistry of root exudates in the rhizosphere by changing the compounds secreted, or by differential adsorption of secreted exudates. These changes may make the rhizosphere less conducive for disease microorganisms and more amenable to protective microorganisms.

vi) Redox active biochar components can participate in a wide range of chemical and biological electron transfer reactions in the rhizosphere, and in this way, influence important processes that depend on electron transfer along the soil-microbe-plant continuum.

vii) Biochar can induce systemic plant defence mechanisms, with elicitors being biochar-borne chemicals, biochar-induced microorganisms, or both.

These different mechanisms are examined below.

3.1.1 Improved Nutrient Supply and Enhanced Plant Growth

Many biochars, particularly those prepared from manures and agricultural residues, contain elements that can directly contribute to plant nutrition (Atkinson et al. 2010). More important than total nutrient content, however, is the proportion of those nutrients that are plant available. While the available proportion decreases with increasing HTT, because nutritive elements become incorporated in poorly-soluble mineral phases or inside the carbon skeleton (Chan and Xu 2009), biochar-born elements may still make a considerable contribution to plant nutrition. For example, on the basis of kinetic curves for nutrient release measured over a range of pH values from 4 to 8, it was estimated that a biochar produced from corn straw by fast pyrolysis at 500°C had the potential to substitute substantial and season-long proportions of P for both extensive and intensive crops (Silber et al. 2010). It was also estimated that the corn straw biochar could be a good initial source of Ca and Mg for plants in acidic soils, and for K in all soil types. The essential role played by plant nutrients for enhancing plant performance is well known.

Besides directly supplying plant nutrients, biochar can improve plant nutrition indirectly via mechanisms that improve nutrient availability. These can include improved retention of nutrients via cation exchange (Taghizadeh-Toosi et al. 2012); increases in the pH of acidic soils and accompanying improvements in nutrient availability and decreases in specific element toxicity (Yamato et al. 2006, Steiner et al. 2007, Novak et al. 2009); changes in P and S transformations and turnover (Pietikainen et al. 2000, DeLuca et al. 2009), and enhanced mycorrhizal functioning (Warnock et al. 2007). More details about how biochar may supply nutrients and improve nutrient availability have been discussed in detail in recent studies and reviews (Chan and Xu 2009, DeLuca et al. 2009, Atkinson et al. 2010, Silber et al. 2010), and will not be further enlarged upon here. Paradoxically, well-nourished plants do not always fare well against foliar and soil pathogens. This is because infection by some pathogens is promoted on plants tissues having high nitrogen contents. For instance, higher leaf N resulted in an increased barley grain yield, but also in an increase in the level of powdery mildew infestation (Graham 1983).

3.1.2 Increase in Beneficial Soil Microorganisms

Biochar is initially sterile following its production, and thus lacks an indigenous population of microorganisms that can potentiate disease suppression, as occurs with suppressive composts (Hoitink et al. 1997). However, when biochar is added to soil it frequently has a profound impact on microbial biomass, microbial community structure, respiration, and enzyme activities in both bulk soil and the rhizosphere (Lehmann et al. 2011). Mixing of biochar into soils frequently stimulates microbial growth and activates dormant soil microorganisms. This results in significant increases in microbial respiration rates in the short time (Steinbeiss et al. 2009, Smith et al. 2010), although respiration rates usually drop following an initial rapid flush. It has been suggested that such biphasic biochar mineralization patterns indicate rapid degradation of labile biochar components followed by slow to negligible degradation of the condensed aromatic ring structures. The labile organic components are residues of volatile and semi-volatile organic compounds created during pyrolysis that condense on the biochar during cooling, and consist of a wide array of molecules including *n*-alkanoic acids, hydroxy and acetoxy acids, amines, benzoic acids, diols, triols, and phenols (Graber et al. 2010, Spokas et al. 2011). Other aspects of biochar addition to soil that have been suggested to influence microbial populations and biodiversity include the possibility that the biochar microporous carbon skeleton provides a physical refuge for some microorganisms (Warnock et al. 2007), and that production of ethylene in the root zone of biochar-amended soils reduces nitrifying bacteria (Spokas et al. 2010). While there is still much to learn about the mechanisms by which biochar affects microbial abundance and community structure, it is well-known that soil microorganisms can have a tremendous impact on both plant productivity and plant resistance to disease.

It has been often observed that addition of biochar to soil results in significant augmentation of symbiotic interactions between plants and AMF (Warnock et al. 2007). AMF, which are obligate symbiotic soil fungi that colonize the roots of vascular plants, make up a major family of soil microorganisms known for its positive impact on plant productivity and health. Plant-AMF symbiosis frequently results in increased plant uptake of immobile nutrients, principally phosphorus, from the soil (Harrison 1999), resulting in enhanced plant vigor. The symbiosis also frequently increases plant resistance to drought and improved pest tolerance (Nelsen and Safir 1982). AMF–host symbiosis also results in modulation of PR proteins and phytohormones in the host plant (Shaul et al. 1999, Shaul-Keinan et al. 2002). Warnock et al. (2007) summarized four mechanisms by which biochar may augment mycorrhizal abundance and impact its functioning: (i) by altering nutrient availability or soil physical-chemical properties that affect AMF;

(ii) indirectly through either beneficial or detrimental effects on other soil microbes; (iii) by altering plant-AMF signaling processes or detoxifying allelochemicals that affect AMF colonization; and (iv) by providing physical refuge from fungal grazers. Such mechanisms may also affect other soil dwelling fungi including plant pathogens.

It is known that rhizosphere microorganisms in general, and selected strains belonging to the genera *Pseudomonas*, *Bacillus*, and *Trichoderma* in particular, can enhance plant growth and stimulate plant resistance to disease in many cropping systems (Windham et al. 1986, Kloepper et al. 2004, Mercado-Blanco and Bakker 2007). Intriguingly, it was reported that culturable numbers of *Pseudomonas* spp., *Trichoderma* spp., and *Bacillus* spp., were significantly higher in the rhizosphere and bulk soil of mature pepper plants whose growth (Graber et al. 2010) and resistance to foliar disease (Elad et al. 2010) were enhanced by biochar additions. It was reported that the vast majority of distinct bacterial isolates (16 out of 18 total isolates) shared high sequence identity (98% or more) based on partial 16S rRNA gene analysis, with strains of *Pseudomonas, Mesorhizobium, Brevibacillus*, and *Bacillus* spp. known for their ability to act as either or both plant growth promoting agents or biocontrol agents (Graber et al. 2010). Antibiotic producers (*Pseudomonas aeruginosa* and *P. mendocina* strains) were also identified in biochar-amended soil (Graber et al. 2010). The role of biochar in augmenting beneficial microbes that can elicit plant systemic resistance is further detailed below.

3.1.3 Sorption of Toxins Produced by Disease Pathogens

Soil pathogens face a complex root membrane barrier composed of a variety of biopolymers having different chemical linkages. In order to breakdown this barrier, the pathogen secretes an assortment of enzymes, each of which is active against a specific component of the cell wall. The most important enzymes secreted by pathogens are those capable of degrading cellulose and pectin. In addition, soil pathogens secrete inhibitory compounds such as oxalic, phenylacetic, succinic, and lactic acids. These compounds reduce pH near and in the infected tissues, playing an important role in pathogenesis. The adsorbent quality of biochars could potentially play a protective role against pathogen attack by adsorbing secreted enzymes and inhibitory compounds, thus reducing their contact with plant root cell walls. Biochar, frequently being alkaline and having a high buffer capacity, can also moderate the affect of these acids on pH near the root.

Little is known about the potential for different biochars to adsorb enzymes. Activated carbon, having a large pore volume and specific surface area, as well as a significant fraction of pores in a size range suitable for enzyme adsorption (300–1000 angstrom), has a substantial adsorption capacity for enzymes. For example, the Langmuir capacity of a commercial activated carbon (SSA 1073 m^2/g) for cellulase was about 1565 mg/g at 30°C (Boukraa-Oulad et al. 2010). It is expected that biochars, having significantly lower SSAs than activated carbon and far more irregular pore structures, will have lower adsorption capacities; however, there is not much data about this in the literature. Immobilization due to adsorption of cellulase on activated carbon did not reduce its activity; to the contrary, immobilization on the carbon support reduced the enzyme inhibition that commonly accompanies hydrolysis of cellulose (Daoud et al. 2010). In a different experiment, both the enzyme (β-glucosidase) and substrate (cellubiose) were so strongly adsorbed to activated carbon that there was no resultant enzyme activity against the substrate (Lammirato et al. 2011). Low SSA wood biochar, however, adsorbed β-glucosidase but not cellubiose, and its activity was reduced only by about 30% despite being immobilized by adsorption (Lammirato et al. 2011). This may be suggestive that ultimately, the effect of enzyme immobilization on biochar as a result of adsorption will depend on the availability of the substrate. This mechanism may explain the observation that fast pyrolysis biochar from switchgrass had inconsistent impacts on the activities of four soil enzymes (β-glucosidase, β-N-acetylglucosaminidase, lipase, and leucine aminopeptidase) in different soils (Bailey et al. 2011).

In the case of pectolytic, cellulolytic and other cell wall-degrading enzymes secreted by pathogens, it may transpire that they could retain their potency during adsorption yet be rendered ineffective because of immobilization by the biochar. If the enzymes are not free to come into contact with the root cell walls, biochar could end up protecting underground plant tissue from pathogen attack. This conjecture needs to be underpinned by model-driven enzyme adsorption and activity experiments. It also needs to be examined whether different biochars play a role in adsorbing and detoxifying small molecule toxins produced by disease pathogens. A possible indication of this was reported by Elmer and Pignatello (2011), who showed that biochar additions enhanced the colonization of asparagus roots by AMF despite the addition of allelopathic agents known to reduce AMF colonization in asparagus. This was attributed to the adsorption affinity of the biochar for the allelopathic compounds, reinforcing earlier work by Wardle et al. (1998). While activated carbons are known to adsorb dicarboxylic acids (Lee et al. 1986), it still needs to be seen if non-activated charcoals have any appreciable adsorption ability for the dicarboxylic acids and other common inhibitors secreted by pathogens.

3.1.4 Suppression of Soil-borne Pathogens by Biochar-Derived Toxic Organic Compounds

Biochars contain residual tars which are composed of a complex mixture of hundreds of individual organic compounds (Schnitzer et al. 2007a, Schnitzer et al. 2007b, Schnitzer et al. 2008, Graber et al. 2010, Spokas et al. 2011), representing numerous chemical classes (alcohols, acids, amines, aldehydes, ketones, phenols, lignin monomers and dimers, sugars, N-heterocyclics, carbocyclics, aliphatics, fatty acids, and furans). Graber et al. (2010) identified a number of biochar compounds known to adversely affect microbial growth and survival, including ethylene and propylene glycol, hydroxy-propionic and butyric acids, benzoic acid and *o*-cresol, quinones (resorcinol and hydroquinone), and 2-phenoxyethanol. Low levels of these toxic compounds could suppress sensitive components of the soil microbiota thereby resulting in proliferation of resistant microbial communities. A possible indication of such a mechanism was the identification of an isolate with 100% 16S rRNA gene sequence identity to the nitrophenol-degrader *Nocardioides nitrophenolicus* in biochar-amended soil (Graber et al. 2010). Microorganisms which excel at degrading toxic organic contaminants generally are more resistant to a variety of toxic organic compounds. Moreover, several antibiotic producers such as the fluorescent pseudomonads (including the *P. aeruginosa* and *P. mendocina* strains) were detected in the rhizosphere of the biochar-amended pepper plants (Graber et al. 2010). It is known that microbial antibiotic producers can be antagonistic towards soil pathogens.

3.1.5 Adsorption of Root Exudates

Roots typically release large quantities of various exudates involving an extensive array of organic and inorganic chemicals (Bais et al. 2006). These include high molecular weight polysaccharides and enzymes, intermediate weight compounds (flavonoids, fatty acids, tannins, carbohydrates, steroids, terpenoids and vitamins), as well as low molecular weight sugars, amino acids, organic acids, and phenolic compounds. These compounds have an important influence on the growth and development of soil microorganisms. Given the well-known adsorption ability of biochar for small and large organic molecules (e.g., Bornemann et al. 2007, Chen and Chen 2009, Joseph et al. 2010, Graber et al. 2011a, Graber et al. 2011b), which is generally governed by interaction affinities and overall site capacity, it is expected that biochar will adsorb various root exudate chemicals, changing the chemistry of the rhizosphere and altering the microbial community structure. Conceivably, this could make the rhizosphere less conducive to the development of disease microorganisms.

There is virtually no information in the literature concerning the adsorption of common root exudate chemicals by biochar. Addition of activated charcoal to a nutrient solution for the hydroponic culture of tomato resulted in both a considerable decrease in the carbon concentration in the solution and significant increases in the dry weight of plant and in fruit yield (Yu et al. 1993). It was concluded that the growth of tomato in hydroponic culture was inhibited by root exudates which were removed from the nutrient solution by adsorption on the activated charcoal. In principle, biochar could also change the nature of the exudates secreted by the plant, but such detailed examinations of biochar impacts on plant physiology have yet to be carried out.

3.1.6 Mediation of Redox Processes

A number of the organic constituents making up the labile fraction of biochars (Graber et al. 2010) are the same as those identified in natural Dissolved Organic Matter (DOM) as being redox active (Fimmen et al. 2007). This includes small molecules such as quinones, hydroxylated benzenes and benzoic acid moieties, and other redox-active aromatics, as well as large humic-like macromolecules (Fimmen et al. 2007). In natural DOM, these components are known to be important intermediaries in microbial metabolic processes and to facilitate the biological cycling of metals with multi-oxidation states (Lovley et al. 1998). As a consequence, DOM has been shown to influence both microbial ecology and function (Visser 1985). Moreover, redox active components in DOM can participate in abiotic electron transfer processes (Fimmen et al. 2007). Quinoid groups promote chemical reduction of Fe and Mn oxides, while carboxylate moieties accelerate chemical oxidation. Carboxylate moieties can also interact with metals having different oxidation states to form stable metal-organic ligands, thus increasing their aqueous concentrations.

Graber et al. (unpublished) compared reducing power of aqueous extracts of biochars produced from 3 feedstocks (wood, olive pomice, greenhouse waste) at different temperatures to that of natural DOM from various sources (Suwannee River DOM and fulvic acid; Pahokee peat fulvic and humic acids). On a unit molar OM basis, the biochar extracts had reducing powers comparable to those of the natural DOMs. Furthermore, biochar aqueous extracts caused significant solubilization of Mn from 4 different soils (Graber et al. unpublished). Accordingly, soluble biochar organic chemicals may participate in a wide range of chemical and biological redox and metal complexation reactions in the rhizosphere, and in this way, influence important processes along the soil-microbe-plant continuum. Such processes include microbial ecology and function, nutrient cycling, root uptake of nutrients, and free radical scavenging. All of these can have significant impacts on plant responses to disease.

The biochar carbon skeleton itself may also participate in redox reactions in the rhizosphere (Joseph et al. 2010). The biochar aromatic C structure functions mainly as a reducing agent, with O_2 being the dominant electron acceptor. In the absence of O_2, alternative electron acceptors (e.g., MnOOH, MnO_2) may also oxidize biochar surfaces (Nguyen and Lehmann 2009). Most low temperature biochars are semiconductors and have a variety of minerals including Mn and Fe in various phases at surfaces (Joseph et al. 2010). Sweep voltammetry analysis of several biochars showed that low temperature and high mineral content biochars have high redox activities. SEM analysis confirmed that during both oxidation and reduction cycles, there were substantial changes in the physical surface structure of those biochars. Thus the carbon skeleton itself can play a role in affecting both local soil redox reactions and microbial electron transfer reactions, and it has been suggested there could be a coupling between these reactions and the transfer of cations, anions, protons and electrons in the plant when root hairs or roots become attached to the biochar (Joseph et al. 2010).

3.1.7 Induction of Systemic Plant Defense Mechanisms

Chemical Elicitors. In aqueous solution, some elements (for example, Si) can directly elicit SAR (Hammerschmidt 2011), and others indirectly by inducing abiotic stresses such as osmotic or hydroxide stress, which then stimulate resistance pathways in plants (Wiese et al. 2004, Wiese et al. 2005). Seeing that many biochars contain a wealth of water soluble minerals, it is possible that these mineral phases contribute to systemic resistance responses (Graber et al. 2010). Minerals commonly found in biochars include various Ca phosphates and nitrates, sylvite (KCl), NaCl, quartz and amorphous silicon dioxide, calcite ($CaCO_3$), and minor phases such as anhydrite ($CaSO_4$) and oxides and hydroxides of Ca, Mg, Al, Ti, Zn, Mn and Fe (Amonette and Joseph 2009). Dissolution of these minerals into the soil solution can lead to high salt concentrations (causing osmotic stresses), high pH (hydroxide stress), or to high Si concentrations, all of which can stimulate induced resistance.

Graber et al. (2010) also conjectured that plants grown in biochar-amended soils could exhibit a systemic response to low levels of biochar-borne phytotoxic organic compounds. There is evidence that low doses of organic chemicals can stimulate resistance to plant diseases (Calabrese and Blain 2009), while the same chemicals at higher doses could have a phytotoxic effect. Hormesis, defined as low dose stimulation/high dose inhibition by chemicals, operates in a variety of plants as evidenced by numerous plant growth indicators, certain metabolic processes, and incidence of plant disease (Calabrese and Blain 2009). Chemicals from many classes, including phenols, carboxylic and fatty acids, aromatic compounds, hydrocarbons, and others, have been reported to evoke inverted U-shaped hermetic dose-response curves in plants (Calabrese and Blain 2009). This effect has also been documented for various allelochemicals (Liu et al. 2003), the plant hormone ethylene (Pierik et al. 2006), various herbicides (Velini et al. 2008, Cedergreen et al. 2009), and some antibiotics (Migliore et al. 2010). Biochar amendment has been reported to cause ethylene production in certain soils (Spokas et al. 2010).

Biological Elicitors. Microorganisms in general, and specifically selected strains belonging to the genera *Pseudomonas*, *Bacillus*, and *Trichoderma*, are known for their ability both to improve plant growth and to potentiate plant systemic resistance against diseases and pests in many plant species (Koike et al. 2001, Srinath et al. 2003, Kloepper et al. 2004, Gravel et al. 2007, Kaewchai et al. 2009). Analysis of pepper root-associated bacterial diversity in greenhouse experiments using culture-independent high-throughput 16S rRNA pyrosequencing showed significantly higher relative amounts of the *Bacteroidetes* phylum and a decrease in *Proteobacteria* in biochar amended potting mixtures compared to control plants (Kolton et al. 2011). On the genus level, a substantial increase in abundance of *Flavobacterium* as well as a more modest induction of the genera *Algoriphagus, Cellvibrio, Nitrospira, Chitinophaga* and *Hydrogenophaga* were seen in biochar amended soils (Fig. 2). Many members of the *Flavobacterium* genus, which is widely distributed in nature, produce a wide range of antibiotics (Clark et al. 2009), and some *Flavobacterium* isolates have biocontrol capabilities (Hebbar et al. 1991, Alexander and Stewart 2001, Gunasinghe and Karunaratne 2009). In addition, some flavobacteria strains have been shown to elicit plant systemic responses to different diseases (An et al. 2008, An et al. 2009). Members of the *Flavobacterium* genus commonly possess an arsenal of extracellular enzymes such as proteinases and chitinases which enable them to degrade bacteria, fungi, insects and nematode constituents (Bernardet and Bowman 2006). Other hydrolytic enzyme-producing genera including *Chitinophaga* (Bacteroidetes) and *Cellvibrio* (Betaproteobacteria) were also induced in the rhizosphere of the biochar-amended pepper plants (Kolton et al. 2011). Such biopolymer-degrading bacteria may release chitin oligomers, well known elicitors of ISR (Rajkumar et al. 2008a), to the soil environment. Antibiotic producing *P. mendocina* and *P. aeruginosa* strains were isolated from the rhizosphere of the biochar-amended treatments (Graber et al. 2010); similar strains are known to promote plant growth and induce systemic

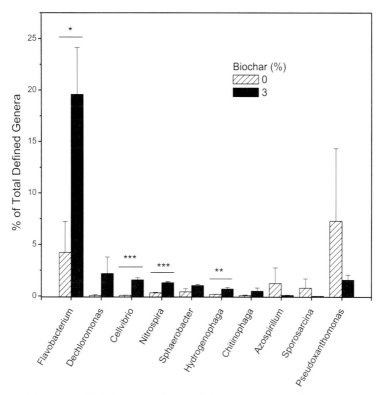

Figure 2. The effect of biochar amendment of the root-associated bacterial community composition of mature sweet pepper (*Capsicum annuum* L.) plants was evaluated using molecular techniques (Kolton et al. 2011). Molecular fingerprinting (denaturing gradient gel electrophoresis and terminal restriction fragment length polymorphism) of 16S rRNA gene fragments showed a clear differentiation between the root associated bacterial community structures of biochar-amended and control plants. The figure shows the relative abundance of root-associated bacterial genera in the control and biochar-amended soils identified using the pyrosequencing of bar-coded 16S rRNA amplicons. Asterisks indicate significant differences in the relative abundance of groups in the biochar versus control samples using the Mann-Whitney test at a confidence level of 94.2 (*), 95 (**) and 99% (***). Reprinted with permission from Kolton et al. (2011).

resistance (Haas and Défago 2005, Vespermann et al. 2007, Mazurier et al. 2009). Hence, it appears that biochar somehow selects for a more beneficial microbial community.

3.2 Biochar Interaction with Soil Pesticides

In a chapter addressing how the addition of biochar to soil affects plant resistance to disease and soil suppressiveness, it is important to note that biochar could have a negative impact on the efficacy of soil-applied pest products, including fungicides, insecticides, and herbicides. This is because many biochars have a high adsorption affinity and capacity for a variety of organic compounds. For example, a progressive increase in adsorption with increasing biochar content was demonstrated for the fungicide pyrimethanil (Yu et al. 2010). Biochars having high SSAs can be particularly challenging for pest control (Graber et al. 2012), since for many compounds, their adsorption strength is commonly much greater than that of low SSA biochars (Bornemann et al. 2007, Chen and Chen 2009, Wang et al. 2010, Yang et al. 2010). Biochar SSA generally increases with increasing pyrolysis temperature.

Because biochar may be an excellent adsorbent for various pest management products, larger product doses may be required to obtain adequate pest control, for example, as seen with 1,3-dichloropropene against nematodes (Graber et al. 2011b) (Fig. 3). Depending on the rate of biochar addition and the biochar SSA, it is possible that needed doses would exceed

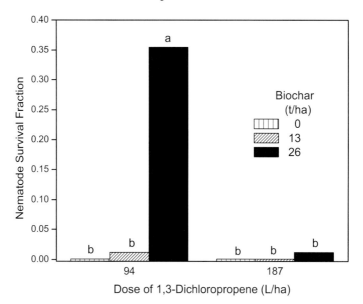

Figure 3. Efficacy of the fumigant 1,3-dichloropropene (DCP) against nematodes was tested in soil amended with 0, 13 or 26 t/ha of biochar produced by fast pyrolysis at 500°C from corn straw. The biochar had a low surface area (SSA) of 3 m²/g. Three doses of DCP were tested: 0, 94 and 187 L/ha. The manufacturer's suggested application rate range for DCP is 85-514 L/ha. Results are shown for 94 and 187 L/ha DCP in terms of fraction of nematode survival at the three biochar amendment rates compared to survival with no DCP and no biochar. From the results, it is clear that at 26 t/ha biochar, twice as much DCP (187 versus 94 L/ha) had to be applied to obtain good nematode control. However, the required dose (187 L/ha) was well within the manufacturer's suggested range. It was calculated that for a high SSA biochar, nematode control would have failed at even the highest recommended fumigant dose. Data taken from Graber et al. (2011b) with permission.

maximum label rates, as reported for two herbicides when high SSA biochar was added to soil (Graber et al. 2012). Alternatively, the soluble organic component of biochar may form complexes with soil-applied herbicides (Cabrera et al. 2011), thus enhancing their downward transport through the soil (Graber et al. 1995, Cabrera et al. 2011). In general, the outcome of pest management under biochar application will depend on the rate of biochar addition and on its physical and chemical characteristics, such as SSA. Other biochar attributes that may influence absorption ability and hence pesticide efficacy include CEC, particularly for cationic substances, and oxidation status of biochar surface functional groups (Borisover and Graber 2002). Based on information available to now, low SSA (<10 m^2/g) biochars added at rates of up to 20 t/ha appear to present low risks for pest management, but this observation needs to be shored up with additional experiments.

3.3 The Known Unknowns and the Unknown Unknowns

While a number of potential mechanisms for the impact of biochar on development and progression of plant disease are listed and discussed above, there is no certainty that they, or others, are indeed responsible singly or together. There is no doubt that the listing is yet incomplete, and that there are other possible mechanisms. The following is a non-exhaustive list of known unknowns regarding the effect of biochars on:

- the plant metabolome both in aerial and below ground parts
- plant hormones and hormone responses
- root development and architecture
- plant secretion of root exudates
- exudate composition in the soil
- soil chemical and physical characteristics
- activities and secretions of soil microorganisms
- combined and synergistic relationships between different factors

Biochar-induced changes in any of the above could also influence the development and progression of plant disease, yet nothing is known about them. Moreover, there is no information on how these impacts may change over time and how temporal plant responses are affected. It remains to be seen how biochar production parameters and feedstocks affect plant responses, and to determine in which cropping systems biochar provides protection, and against what types of diseases. To move towards conceptual understanding, model-driven, mechanistic studies aimed at comprehending the mechanisms involved in disease reduction are needed. Future research should address the nature of the induced resistance pathways, and should also examine whether biochar helps to prime plants against abiotic stresses.

Additionally, there is a need to address potential risks to pest control associated with biochar use, and to determine the best methods of using biochar in agricultural systems.

Improved plant health is a further benefit that may be provided by biochar addition to soil, together with increased crop yield, enhanced soil fertility and improved soil structure. The development of a variety of agricultural markets for biochar products, including as a disease control agent, will play an essential role in the ultimate adoption of the biomass pyrolysis/biochar platform as a tool for mitigation of, and adaptation to, climate change.

Acknowledgements

Biochar research at the Agricultural Research Organization, Volcani Center, Israel is supported by grants from the Chief Scientist of the Israel Ministry of Agriculture and Rural Development, projects 301-0693-10 and 261-0848-11, the Autonomous Province of Trento, Call for Proposal Major Projects 2006, Project ENVIROCHANGE, and the Israel-Italy Program 2011, project number 301-0750-11. This chapter is contribution no. 601/12 of the Agricultural Research Organization, The Volcani Center, Israel. The authors appreciate the critical review by Dr. Eddie Cytryn. The authors would also like to acknowledge the assistance of Menachem Borenshtein, Eyal Cohen, Eddie Cytryn, Omer Frenkel, Alex Furman, Zeraye Mehari Haile, Amit Kumar Jaiswal, Max Kolton, Irit Levkovitch, Beni Lew, Yael Meller-Harel, Dalia Rav-David, Sergey Segal, Ran Shulhani, Avner Silber, Ludmilla Tsechansky, and Haguy Yasour in their various biochar endeavors.

Glossary

E	:	Ethylene
DOM	:	Dissolved Organic Matter
HTT	:	Highest Treatment Temperature
IR	:	Induced Resistance
ISR	:	Induced Systemic Resistance
JA	:	Jasmonic Acid
PR	:	Pathogenesis-Related
PGPR	:	Plant Growth-Promoting Rhizobacteria
PGPF	:	Plant Growth-Promoting Fungi
SA	:	Salicylic Acid
SAR	:	Systemic Acquired Resistance
SSA	:	Specific Surface Area

References

Ahn, I.P., S.W. Lee, and S.C. Suh. 2007. Rhizobacteria-induced priming in Arabidopsis is dependent on ethylene, jasmonic acid, and NPR1. Mol. Plant-Microbe Interact. 20: 759–768.

Alexander, B.J.R. and A. Stewart. 2001. Glasshouse screening for biological control agents of *Phytophthora cactorum* on apple (*Malus domestica*). NZ J. Crop Hort. Sci. 29: 159–169.

Allen, R.I. 1847. A Brief Compend of American Agriculture C.M. Saxton, New York.

Amonette, J.E. and S. Joseph. Characteristics of biochar: Microchemical properties, p. 33–52. *In:* J. Lehmann and S. Joseph (eds.). 2009. Biochar for Environmental Management: Science and Technology. ed. Earthscan, London.

An, O.D., G.L. Zhang, H.T. Wu, Z.C. Zhang, G.S. Zheng, Y.L. Zhang, X.Z. Li, and Y. Murata. 2008. Properties of an alginate-degrading *Flavobacterium* sp. strain LXA isolated from rotting algae from coastal China. Can. J. Microbiol. 54: 314–320.

An, Q.D., G.L. Zhang, H.T. Wu, Z.C. Zhang, G.S. Zheng, L. Luan, Y. Murata, and X. Li. 2009. Alginate-deriving oligosaccharide production by alginase from newly isolated *Flavobacterium* sp LXA and its potential application in protection against pathogens. J. App. Microbiol. 106: 161–170.

Atkinson, C.J., J.D. Fitzgerald, and N.A. Hipps. 2010. Potential mechanisms for achieving agricultural benefits from biochar application to temperate soils: a review. Plant Soil 337: 1–18.

Ayres, P.G. 1984. The interaction between environmental-stress injury and biotic disease physiology. Annu. Rev. Phytopathol. 22: 53–75.

Aziz, A., P. Trotel-Aziz, L. Dhuicq, P. Jeandet, M. Couderchet, and G. Vernet. 2006. Chitosan oligomers and copper sulfate induce grapevine defense reactions and resistance to gray mold and downy mildew. Phytopathology 96: 1188–1194.

Bailey, V.L., S.J. Fansler, J.L. Smith, and H.J. Bolton. 2011. Reconciling apparent variability in effects of biochar amendment on soil enzyme activities by assay optimization. Soil Biol. Biochem. 43: 296–301.

Bais, H.P., T.L. Weir, L.G. Perry, S. Gilroy, and J.M. Vivanco. 2006. The role of root exudates in rhizosphere interations with plants and other organisms. Annu. Rev. Plant Biol. 57: 233–266.

Belhadj, A., C. Saigne, N. Telef, S. Cluzet, J. Bouscaut, M.F. Corio-Costet, and J.M. Mérillon. 2006. Methyl jasmonate induces defense responses in grapevine and triggers protection against *Erysiphe necator*. J. Agric. Food Chem. 54: 9119–9125.

Bernardet, J.F. and J.P. Bowman. The genus *Flavobacterium*. p. 481–531. *In:* M. Dworkin, S. Falkow, E. Rosenberg, K.-H. Schleifer, and E. Stackebrant (eds.). 2006. The Prokaryotes: a Handbook on the Biology of Bacteria. 3 ed. Springer, New York, NY.

Borisover, M. and E.R. Graber. 2002. Simplified link solvation model (LSM) for sorption in natural organic matter. Langmuir 18: 4775–4782.

Bornemann, L.C., R.S. Kookana, and G. Welp. 2007. Differential sorption behaviour of aromatic hydrocarbons on charcoals prepared at different temperatures from grass and wood. Chemosphere 67: 1033–1042.

Cabrera, A., L. Cox, K.A. Spokas, R. Celis, M.C. Hermosin, J. Cornejo, and W.C. Koskinen. 2011. Comparative sorption and leaching study of the herbicides fluometuron and 4-chloro-2-methylphenoxyacetic acid (MCPA) in a soil amended with biochars and other sorbents. J. Agric. Food Chem. 59: 12550–12560.

Calabrese, E.J. and R.B. Blain. 2009. Hormesis and plant biology. Environ. Poll. 157: 42–48.

Cedergreen, N., C. Felby, J.R. Porter, and J.C. Streibig. 2009. Chemical stress can increase crop yield. Field Crops Res. 114: 54–57.

Chan, K.Y. and Z. Xu. Biochar: Nutrient properties and their enhancement, p. 67–84. *In:* J. Lehmann and S. Joseph (eds.). 2009. Biochar for Environmental Management: Science and Technology. ed. Earthscan, London.

Chen, B.L. and Z.M. Chen. 2009. Sorption of naphthalene and 1-naphthol by biochars of orange peels with different pyrolytic temperatures. Chemosphere 76: 127–133.

Clark, S.E., B.A. Jude, G.R. Danner, and F.A. Fekete. 2009. Identification of a multidrug efflux pump in *Flavobacterium johnsoniae*. Veterinary Res. 40: 55.

Conrath, U., C.M.J. Pieterse, and B. Mauch-Mani. 2002. Priming in plant-pathogen interactions. Trends Plant Sci. 7: 210–216.

Conrath, U., G.J. Beckers, V. Flors, P. Garcia-Agustin, G. Jakab, F. Mauch, M.A. Newman, C.M.J. Pieterse, B. Poinssot, M.J. Pozo, A. Pugin, U. Schaffrath, J. Ton, D. Wendehenne, L. Zimmerli, and B. Mauch-Mani. 2006. Priming: Getting ready for battle. Mol. Plant-Microbe Interact. 19: 1062–1071.

Daoud, F.B.-O., S. Kaddour, and T. Sadoun. 2010. Adsorption of cellulase *Aspergillus niger* on a commercial activated carbon: Kinetics and equilibrium studies. Colloids and Surfaces B: Biointerfaces 75: 93–99.

DeLuca, T.H., M.D. MacKenzie, and M.J. Gundale. Biochar effects on soil nutrient transformations, p. 251–270. *In:* J. Lehmann and S. Joseph (eds.). 2009. Biochar for Environmental Management: Science and Technology. ed. Earthscan, London.

Downie, A., A. Crosky, and P. Munroe. Physical properties of biochar, p. 13–32. *In:* J. Lehmann and S. Joseph (eds.). 2009. Biochar for Environmental Management: Science and Technology. ed. Earthscan, London.

Elad, Y., D. Rav David, Y. Meller Harel, M. Borenshtein, H. Ben Kalifa, A. Silber, and E.R. Graber. 2010. Induction of systemic resistance in plants by biochar, a soil-applied carbon sequestering agent. Phytopathology 100: 913–921.

Elad, Y., E. Cytryn, Y. Meller Harel, B. Lew, and E.R. Graber. 2011. The biochar effect: Plant resistance to biotic stresses. Phytopathologia Mediterranea 50: 335–349.

Elmer, W.H. and J.J. Pignatello. 2011. Effect of biochar amendments on mycorrhizal associations and fusarium crown and root rot of asparagus in replant soils. Plant Dis. 95: 960–966.

Fimmen, R.L., R.M. Cory, Y.P. Chin, T.D. Trouts, and D.M. McKnight. 2007. Probing the oxidation-reduction properties of terrestrially and microbially derived dissolved organic matter. Geochim. Cosmochim. Acta 71: 3003–3015.

Gaskin, J.W., C. Steiner, K. Harris, K.C. Das, and B. Bibens. 2008. Effect of low-temperature pyrolysis conditions on biochar for agricultural use. Trans. ASABE 51: 2061–2069.

Graber, E.R., Z. Gerstl, E. Fischer, and U. Mingelgrin. 1995. Enhanced transport of atrazine under irrigation with effluent. Soil Sci. Soc. Am. J. 59: 1513–1519.

Graber, E.R., L. Tschansky, Z. Gerstl, and B. Lew. 2012. High surface area biochar negatively impacts herbicide efficacy. Plant Soil 353: 95–106.

Graber, E.R., L. Tschansky, J. Khanukov, and Y. Oka. 2011b. Sorption, volatilization and efficacy of the fumigant 1,3-dichloropropene in a biochar-amended soil. Soil Sci. Soc. Am. J. 75: 1365–1373.

Graber, E.R., Y. Meller-Harel, M. Kolton, E. Cytryn, A. Silber, D. Rav David, L. Tschansky, M. Borenshtein, and Y. Elad. 2010. Biochar impact on development and productivity of pepper and tomato grown in fertigated soilless media. Plant Soil 337: 481–496.

Graham, R.D. 1983. Effects of nutrient stress on susceptibility of plants to disease with particular reference to the trace elements. Adv. Bot. Res. 10: 221–276.

Gravel, V., H. Antoun, and R.J. Tweddell. 2007. Growth stimulation and fruit yield improvement of greenhouse tomato plants by inoculation with *Pseudomonas putida* or *Trichoderma atroviride*: Possible role of indole acetic acid (IAA). Soil Biol. Biochem. 39: 1968–1977.

Gunasinghe, W.K.R.N. and A.M. Karunaratne. 2009. Interactions of *Colletotrichum musae* and *Lasiodiplodia theobromae* and their biocontrol by *Pantoea agglomerans* and *Flavobacterium* sp. in expression of crown rot of "Embul" banana. Biocontrol 54: 587–596.

Haas, D. and G. Défago. 2005. Biological control of soil-borne pathogens by fluorescent pseudomonads. Nat. Rev. Microbiol. 3: 307–319.

Hamiduzzaman, M.M., G. Jakab, L. Barnavon, J.M. Neuhaus, and B. Mauch-Mani. 2005. β-aminobutyric acid-induced resistance against downy mildew in grapevine acts through the potentiation of callose formation and jasmonic acid signaling. Mol. Plant-Microbe Interact. 18: 819–829.

Hammerschmidt, R. 2011. More on silicon-induced resistance. Physiol. Mol. Plant Pathol. 75: 81–82.

Harman, G.E., C.R. Howell, R.A. Vitebro, I. Chet, and M. Lorito. 2004. *Trichoderma* species—opportunistic, avirulent plant symbionts. Nat. Rev. Microbiol. 2: 43–56.

Harrison, M.J. 1999. Molecular and cellular aspects of the arbuscular mycorrhizal symbiosis. Annu. Rev. Plant Physiol. Plant Mol. Biol. 50: 361–189.

Hebbar, P., O. Berge, T. Heulin, and S.P. Singh. 1991. Bacterial antagonists of sunflower (*Helianthus-Annuus* L) fungal pathogens. Plant Soil 133: 131–140.

Hoitink, H.A.J., A.G. Stone, and D.Y. Han. 1997. Suppression of plant diseases by composts. Hortscience 32: 184–187.

Iriti, M., M. Rossoni, M. Borgo, and F. Faoro. 2004. Benzothiadiazole enhances resveratrol and anthocyanin biosynthesis in grapevine, meanwhile improving resistance to *Botrytis cinerea*. J. Agric. Food Chem. 52: 4406–4413.

Joseph, S.D., M. Camps-Arbestain, Y. Lin, P. Munroe, C.H. Chia, J. Hook, L. van Zwieten, S. Kimber, A. Cowie, B.P. Singh, J. Lehmann, N. Foidl, R.J. Smernik, and J.E. Amonette. 2010. An investigation into the reactions of biochar in soil. Austral. J. Soil Res. 48: 501–515.

Kaewchai, S., K. Soytong, and K.D. Hyde. 2009. Mycofungicides and fungal biofertilizers. Fungal Diversity 38: 25–50.

Keiluweit, M., P.S. Nico, M.G. Johnson, and M. Kleber. 2010. Dynamic molecular structure of plant biomass-derived black carbon (biochar). Environ. Sci. Technol. 44: 1247–1253.

Kloepper, J.W., C.M. Ruy, and S. Zhang. 2004. Induced systemic resistance and promotion of plant growth by *Bacillus* spp. Phytopathology 94: 1259–1266.

Koike, N., M. Hyakumachi, K. Kageyama, S. Tsuyumu, and N. Doke. 2001. Induction of systemic resistance in cucumber against several diseases by plant growth-promoting fungi: Lignification and superoxide generation. Eur. J. Plant Pathol. 107: 523–533.

Kolton, M., Y. Meller Harel, Z. Pasternak, E.R. Graber, Y. Elad, and E. Cytryn. 2011. Impact of biochar application to soil on the root-associated bacterial community structure of fully developed greenhouse pepper plants. Appl. Environ. Microbiol. 77: 4924–4930.

Krull, E., J.A. Baldock, J. Skjemstad, and R. Smernik. Characteristics of biochar: Organo-chemical properties, p. 53–66. *In:* J. Lehmann and S. Joseph (eds.). 2009. Biochar for Environmental Management: Science and Technology. ed. Earthscan, London.

Laird, D.A. 2008. The charcoal vision: A win-win-win scenario for simultaneously producing bioenergy, permanently sequestering carbon, while improving soil and water quality. Agron. J. 100: 178–181.

Lammirato, C., A. Miltner, and M. Kaestner. 2011. Effects of wood char and activated carbon on the hydrolysis of cellobiose by b-glucosidase from *Aspergillus niger*. Soil Biol. Biochem. 43: 1936–1942.

Lee, C.-Y.C., E.O. Pedram, and A.L. Hines. 1986. Adsorption of oxalic, malonic, and succinic acids on activated carbon. J. Chem. Engineer. Data 31: 133–136.

Lehmann, J., M.C. Rillig, J. Thies, C.A. Masiello, W.C. Hockaday, and D. Crowley. 2011. Biochar effects on soil biota—a review. Soil Biol. Biochem. 43: 1812–1836.

Liu, D.L., M. An, I.R. Johnson, and J.V. Lovett. 2003. Mathematical modeling of allelopathy. III. A model for curve-fitting allelochemical dose responses. Nonlinearity Biol. Toxicol. Med. 1: 37–50.

Lovley, D.R., J.L. Fraga, E.L. Blunt-Harris, L.A. Hayes, E.J.P. Phillips, and J.D. Coates. 1998. Humic substances as a mediator for microbially catalyzed metal reduction. Acta Hydrochim. Hydrobiol. 26: 152–157.

Matsubara, Y., N. Hasegawa, and H. Fukui. 2002. Incidence of Fusarium root rot in asparagus seedlings infected with arbuscular mycorrhizal fungus as affected by several soil amendments. J. Jpn. Soc. Hort. Sci. 71: 370–374.

Mazurier, S., T. Corberand, P. Lemanceau, and J.M. Raaijmakers. 2009. Phenazine antibiotics produced by fluorescent pseudomonads contribute to natural soil suppressiveness to Fusarium wilt. ISME J. 3: 977–991.

Meller Harel, Y., Y. Elad, D. Rav-David, M. Borenshtein, R. Schulcani, B. Lew, and E.R. Graber. 2012. Biochar mediates systemic response of strawberry to foliar fungal pathogens. Plant Soil Plant and Soil 357: 245–257.

Meller Harel, Y., M. Kolton, Y. Elad, D. Rav-David, E. Cytrin, D. Ezra, M. Borenstein, R. Shulchani, and E.R. Graber. 2011. Induced systemic resistance in strawberry (*Fragaria* × *ananassa*) to powdery mildew using various control agents. IOBC/WPRS Bull. 71: 47–51.

Mercado-Blanco, J. and P.A.H.M. Bakker. 2007. Interactions between plants and beneficial Pseudomonas spp.: exploiting bacterial traits for crop protection. Antonie van Leeuwenhoek 92: 367–389.

Migliore, L., A. Rotini, N.L. Cerioli, S. Cozzolino, and M. Fiori. 2010. Phytotoxic antibiotic sulfadimethoxine elicts a complex hormetic response in the weed *Lythrum Salicaria* L. Dose-Response 8.

Nelsen, C.E. and G.R. Safir. 1982. Increased drought tolerance of mycorrhizal onion plants caused by improved phosphorus nutrition. Planta 154: 407–413.

Nguyen, B.T. and J. Lehmann. 2009. Black carbon decomposition under varying water regimes. Org. Geochem. 40: 846–853.

Novak, J.M., W.J. Busscher, D.L. Laird, M. Ahmedna, D.W. Watts, and M.A.S. Niandou. 2009. Impact of biochar amendment on fertility of a southeastern coastal plain soil. Soil Sci. 174: 105–112.

Perazzolli, M., S. Dagostin, A. Ferrari, Y. Elad, and I. Pertot. 2008. Induction of systemic resistance against *Plasmopara viticola* in grapevine by *Trichoderma harzianum* T39 and benzothiadiazole. Biol. Cont. 47: 228–234.

Pierik, R., D. Tholen, H. Poorter, E.J.W. Visser, and L. Voesenek. 2006. The Janus face of ethylene: growth inhibition and stimulation. Trends Plant Sci. 11: 176–183.

Pietikainen, J., O. Kiikkila, and H. Fritze. 2000. Charcoal as a habitat for microbes and its effect on the microbial community of the underlying humus. Oikos 89: 231–242.

Rajkumar, M., K.J. Lee, and H. Freitas. 2008a. Effects of chitin and salicylic acid on biological control activity of *Pseudomonas* spp. against damping off of pepper. SA J. Bot. 74: 268–273.

Rajkumar, M., K.J. Lee, and H. Freitas. 2008b. Effects of chitin and salicylic acid on biological control activity of *Pseudomonas* spp. against damping off of pepper. SA J. Bot. 74: 268–273.

Reuveni, M., V. Agapov, and R. Reuveni. 1995. Induced systemic protection to powdery mildew in cucumber by phosphate and potassium fertilizers: effects of inoculum concentration and post-inoculation treatment. Can. J. Plant Pathol. 17: 247–251.

Schnitzer, M.I., C.M. Monreal, and G. Jandl. 2008. The conversion of chicken manure to bio-oil by fast pyrolysis III. Analyses of chicken manure, bio-oils and char by Py-FIMS and Py-FDMS. J. Environ. Sci. Health Part B-Pesticides Food Contam. Agri. Wastes 43: 81–95.

Schnitzer, M.I., C.M. Monreal, G.A. Facey, and P.B. Fransham. 2007a. The conversion of chicken manure to bio-oil by fast pyrolysis I. Analyses of chicken manure, biooils and char by C-13 and H-1 NMR and FTIR spectrophotometry. Environ. Sci. Health Part B-Pesticides Food Contam. Agri. Wastes 42: 71–77.

Schnitzer, M.I., C.M. Monreal, G. Jandl, P. Leinweber, and P.B. Fransham. 2007b. The conversion of chicken manure to biooil by fast pyrolysis II. Analysis of chicken manure, biooils, and char by curie-point pyrolysis-gas chromatography/mass spectrometry (Cp Py-GC/MS). Environ. Sci. Health Part B-Pesticides Food Contam. Agri. Wastes 42: 79–95.

Shaul-Keinan, O., V. Gadkar, I. Ginzberg, J.M. Grünzweig, I. Chet, Y. Elad, S. Wininger, E. Belausov, Y. Eshed, N. Atzmon, Y. Ben-Tal, and Y. Kapulnik. 2002. Hormone concentration in tobacco roots change during arbuscular mycorrhizal colonization with *Glomus intraradices*. New Phytol. 154: 501–507.

Shaul, O., S. Galili, H. Volpin, I. Ginzberg, Y. Elad, I. Chet, and Y. Kapulnik. 1999. Mycorrhiza-induced changes in disease severity and PR protein expression in tobacco leaves. Mol. Plant-Microbe Inter. 12: 1000–1007.

Silber, A., I. Levkovitch, and E.R. Graber. 2010. pH-Dependent mineral release and surface properties of cornstraw biochar: agronomic implications. Environ. Sci. Technol. 44: 9318–9323.

Smith, J.L., H.P. Collins, and V.L. Bailey. 2010. The effect of young biochar on soil respiration. Soil Biol. Biochem. 42: 2345–2347.

Smith, N.J.H. 1980. Anthrosols and human carrying capacity in Amazonia. Ann. Assoc. Amer. Geog. 70: 553–566.

Sombroek, W.G. 1966. Amazon Soils Centre for Agricultural Publications and Documentation, Wageningen.

Spokas, K.A., J.M. Baker, and D. Reicosky. 2010. Ethylene: potential key for biochar amendment impacts. Plant Soil 333: 443–452.

Spokas, K.A., J.M. Novak, C.E. Stewart, K.B. Cantrell, M. Uchimiya, M.G. DuSaire, and K.S. Ro. 2011. Qualitative analysis of volatile organic compounds on biochar. Chemosphere 85: 869–882.

Srinath, J., D.J. Bagyaraj, and B.N. Satyanarayana. 2003. Enhanced growth and nutrition of micropropagated *Ficus benjamina* to *Glomus mosseae* co-inoculated with *Trichoderma harzianum* and *Bacillus coagulans*. World J. Microbiol. Biotechnol. 19: 69–72.

Steinbeiss, S., G. Gleixner, and M. Antonietti. 2009. Effect of biochar amendment on soil carbon balance and soil microbial activity. Soil Biol. Biochem. 41: 1301–1310.

Steiner, C., W.G. Teixeira, J. Lehmann, T. Nehls, J.L.V. de Macedo, W.E.H. Blum, and W. Zech. 2007. Long term effects of manure, charcoal and mineral fertilization on crop production and fertility on a highly weathered Central Amazonian upland soil. Plant Soil 291: 275–290.

Taghizadeh-Toosi, A., T.J. Clough, R.R. Sherlock, and L.M. Condron. 2012. Biochar adsorbed ammonia is bioavailable. Plant Soil 350: 57–69.

Ton, J. and B. Maunch-Mani. Elucidating pathways controlling induced resistance, p. 99–109. *In:* G. Voss and G. Ramos (eds.). 2003. Chemistry of Crop Protection. ed. Wiley-VCH, Weinheim.

Trouvelot, S., A.L. Varnier, M. Allègre, L. Mercier, F. Baillieul, C. Arnould, V. Gianinazzi-Pearson, O. Klarzynski, J.-M. Joubert, A. Pugin, and X. Daire. 2008. A β-1,3 glucan sulfate induces resistance in grapevine against *Plasmopara viticola* through priming of defense responses, including HR-like cell death. Mol. Plant-Microbe Interact. 21: 232–243.

Vallad, G.E. and R.M. Goodman. 2004. Systemic acquired resistance and induced systemic resistance in conventional agriculture. Crop Sci. 44: 1920–1934.

Van der Ent, S., S.C. Van Wees, and C.M. Pieterse. 2009. Jasmonate signaling in plant interactions with resistance-inducing beneficial microbes. Phytochemistry 70: 1581–1588.

Van Loon, L.C., P.A.H.M. Bakker, and C.M. Pieterse. 1998. Systemic resistance induced by rhizosphere bacteria. Annu. Rev. Phytopathol. 36: 453–483.

Van Wees, S.C.M., C.M.J. Pieterse, A. Trijssenaar, Y.A.M. Vant Westende, F. Hartog, and L.C. Van Loon. 1997. Differential induction of systemic resistance in *Arabidopsis* by biocontrol bacteria. Mol. Plant-Microbe Interact. 10: 716–724.

Velini, E.D., E. Alves, M.C. Godoy, D.K. Meschede, R.T. Souza, and S.O. Duke. 2008. Glyphosate applied at low doses can stimulate plant growth. Pest Manag. Sci. 64: 489–496.

Vespermann, A., M. Kai, and B. Piechulla. 2007. Rhizobacterial volatiles affect the growth of fungi and *Arabidopsis thaliana* Appl. Environ. Microbiol. 73 5639–5641.

Visser, S.A. 1985. Physiological action of humic substances on microbial-cells. Soil Biol. Biochem. 17: 457–462.

Walters, D., D. Walsh, A. Newton, and G. Lyon. 2005. Induced resistance for plant disease control: Maximizing the efficacy of resistance elicitors. Phytopathology 95: 1368–1373.

Wang, H.L., K.D. Lin, Z.N. Hou, B. Richardson, and J. Gan. 2010. Sorption of the herbicide terbuthylazine in two New Zealand forest soils amended with biosolids and biochars. J Soil Sed. 10: 283–289.

Wardle, D.A., O. Zackrisson, and M.C. Nilsson. 1998. The charcoal effect in Boreal forests: mechanisms and ecological consequences. Oecologia 115: 419–426.

Warnock, D.D., J. Lehmann, T.W. Kuyper, and M.C. Rillig. 2007. Mycorrhizal responses to biochar in soil—concepts and mechanisms. Plant Soil 300: 9–20.

Wiese, J., T. Kranz, and S. Schubert. 2004. Induction of pathogen resistance in barley by abiotic stress. Plant Biol. 6: 529–536.

Wiese, J., H. Wiese, J. Schwartz, and S. Schubert. 2005. Osmotic stress and silicon act additively in enhancing pathogen resistance in barley against barley powdery mildew (in English). J. Plant Nutr. Soil Sci. 168: 269–274.

Windham, M.T., Y. Elad, and R. Baker. 1986. A mechanism for increased plant growth induced by *Trichoderma* spp. Phytopathology 76: 518–521.

Yamato, M., Y. Okimori, I.F. Wibowo, S. Anshori, and M. Ogawa. 2006. Effects of the application of charred bark of *Acacia mangium* on the yield of maize, cowpea and peanut, and soil chemical properties in South Sumatra, Indonesia. Soil Sci. Plant Nutr. 52: 489–495.

Yang, X.B., G.G. Ying, P.A. Peng, L. Wang, J.L. Zhao, L.J. Zhang, P. Yuan, and H.P. He. 2010. Influence of Biochars on Plant Uptake and Dissipation of Two Pesticides in an Agricultural Soil. J. Agric. Food Chem. 58: 7915–7921.

Yu, J.Q., K.S. Lee, and Y. Matsui. 1993. Effect of the addition of activated-charcoal to the nutrient solution on the growth of tomato in hydroponic culture. Soil Sci. Plant Nutr. 39: 13–22.

Yu, X.Y., L.G. Pan, G.G. Ying, and R.S. Kookana. 2010. Enhanced and irreversible sorption of pesticide pyrimethanil by soil amended with biochars. J. Environ. Sci. China 22: 615–620.

3

Biochar-Fungi Interactions
in Soils

Katja Wiedner[a,*] and Bruno Glaser[b]

Introduction

Martin-Luther-University Halle-Wittenberg, Soil Biogeochemistry, von-Seckendorff-Platz 3, 06120 Halle, Germany.
[a]E-mail: katja.wiedner@landw.uni-halle.de
[b]E-mail: bruno.glaser@landw.uni-halle.de
*Corresponding author

Introduction

1 Properties of Biochar that may affect Fungal Communities

Biochar is the product of incomplete combustion of biomass produced under low or limited oxygen supply. It is highly diverse regarding its chemical (Czimczik et al. 2002, Schimmelpfennig and Glaser 2012) and physical properties (Downie et al. 2009). Three physico-chemical features of biochar are potentially involved in its ecological impacts (Fig. 1). Poly-condensed aromatic moieties in biochar are resistant against microbial degradation and, therefore, are responsible for its long-term C sequestration potential (Fig. 1, Glaser et al. 2002, Kuzyakov et al. 2009). Functional groups on the edges of biochar can enhance soil quality by providing exchange sites for cationic soil nutrients (Glaser et al. 2002, Glaser and Birk 2012). Last but not least, the porous physical structure of biochar enables a high water holding capacity and enlarges the surface area, leading to a higher physical sorption of dissolved organic molecules and serving as habitat for soil

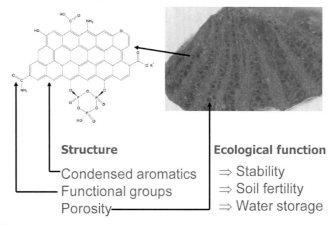

Figure 1. Chemical and physical structure of biochar and its major physico-chemical properties that can have an ecological function.

microorganisms including penetration by fungal hyphae (Fig. 1). Therefore, it is obvious that biochar amendment to soil changes the physical and chemical environment, and therefore have an influence on soil microbial community and activity (e.g., Yin et al. 2000, Kim et al. 2007, O' Neill et al. 2009, Grossman et al. 2010). However, little is known on how soil fungi interact with biochar.

1.1 What are Soil Fungi?

Soil fungi are heterotrophic eukaryotic organisms that differ in morphology, cytology and phylogeny. It is assumed that 500,000 to 10,000,000 species of fungi exist, but only 80,000 fungal species were described so far (Hawksworth 2001). Phylogenetically, they represent a heterogeneous group consisting of real fungi (Kingdom Opisthokonta, Fungi), fungus-like organisms like slime-molds (Phylum Myxomycota, Amoebozoa, Eumycetazoa), and omycota (Phylum Oomycota, Chromalveolata, Stramenopiles) (Rossman and Palm 2006, Ottow 2011, Fig. 2). Fungi are ubiquitous in soils and represent the greatest diversity among soil microorganisms (Giri et al. 2005) belonging to all phylogenetic fungal groups mentioned above and presented in Fig. 2. In terms of their ecological role, fungi can be classified as pathogens, symbionts and saprophytes. Myxomycota feed on soil microorganisms (phagotrophic) and contribute therefore to indirect mineralization processes in soils. Additionally, they are an important food source for organisms like snails, insects or nematodes (Ottow 2011). Zygomycota are fast growing saprophytic fungi dominating the first step of microbial succession in soils. Ascomycota being dominated by Aspergillus and Penicillium live mostly

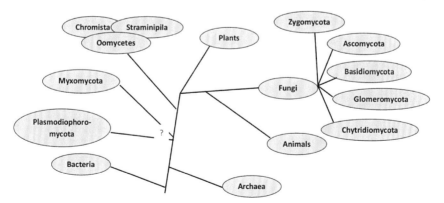

Figure 2. Relationship between the major phylogenetic groups of organisms. Fungal phyla include Zygomycota, Ascomycota, Basidiomycota, Glomeromycota, and Chytridiomycota, while Oomycota, Myxomycota and Plasmodiophoromycota are distinct from fungi. Please note that fungi are more closely related to animals than to plants (modified from Rossman and Palm 2006 and Moore et al. 2011).

saprophytic or phytopathogenic. Basidiomycota vary in ecological ways of life being saprophytic, phytopathogenic or symbiontic.

1.2 Fungal Contribution to Ecosystem Services

The internationally relevant definition for the term 'ecosystem services' was given by the Millennium Ecosystem Assessment (MEA) between 2001 and 2005 and describes the benefits people obtain from ecosystems: *Provisioning services* such as food, water, timber, and fiber; *regulating services* that affect climate, floods, disease, wastes, and water quality; *cultural services* that provide recreational, aesthetic, and spiritual benefits; and *supporting services* such as soil formation, photosynthesis, and nutrient cycling (Hallett et al. 2010, Hoormann 2011).

Soil fungi are involved in many of these services, like nutrient cycling, water dynamics in soil, filtering and bioremediation of soil contaminants or soil aggregation and the production and release of greenhouse gases (Fig. 3, Brussaard 1997, Dighton 2003, Lehmann and Joseph 2009, Hoormann 2011). Well studied examples are arbuscular mycorrhizal (AM) fungi which take

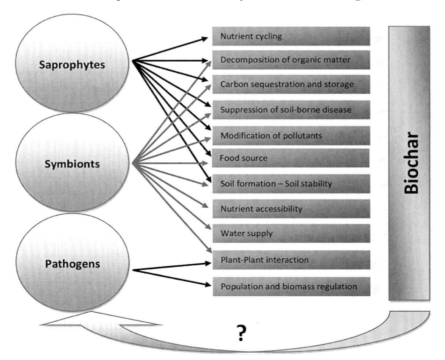

Figure 3. Ecosystem services provided by soil fungi, splitted among ecological groups. Currently, the effect of biochar addition on soil fungi is largely unknown (question mark). Please note that pathogenic fungi cause negative effects on ecosystem services.

over numerous tasks in ecosystems including soil adherence and stability as well as increased water retention (Gianinazzi et al. 2010). Mycorrhizal plants are more resistant against drought, salinity, heavy metals pollution and mineral nutrient depletion. Furthermore, the resistance against biotic stress (e.g., phytopathogens) is enhanced and plant growth is promoted, while fertilizers can be reduced. Recent studies suggest that symbiotic fungi, like mycorrhizal, have the potential to decompose soil C and influence therefore not only the input but also losses of soil C (Talbot et al. 2008).

Saprophytes are important for decomposing and mineralizing organic materials. Besides that, they control nutrient cycling in soil and nutrient availability, and plant community dynamics (Bardgett et al. 2005, Ritz and Young 2009). Six et al. (2006) summarized that C turnover in fungal-dominated communities is slower compared to bacteria. They suggested that fungi incorporate more soil C than bacteria, and the fungal cell walls are more recalcitrant than those of bacteria. Environmental changes could trigger fungal growth, which would cause a release of more CO_2 into the atmosphere through an increased decay of organic matter.

The total biomass of microorganisms in soil amounts only 0.2–4% of total organic carbon but overtakes important functions regarding nutrient and energy flows in soils due to the high turnover rates. Despite the amount of biomass, saprophytic and mycorrhizal fungi occur as sink but also as a source for C, N, P, S and other micronutrients for plants and other microorganisms (Ottow 2011).

The effects of biochar amendment on ecosystem services provided by soil fungi are largely unknown (Fig. 3). Therefore, we reviewed the current knowledge about the effects of Biochar addition on soil fungi.

1.3 Biochar as Habitat for Soil Fungi

The chemical and physical characteristics of biochar play a crucial role on selecting the groups of microorganisms who will colonize its surface and the pores inside biochar particles. Physical properties, like the porous structure, depend on feedstock material and process pressure, whereas the pore space and surface area is the result of feedstock material and charring temperature (Fig. 4). Pyrolysis at high temperature generates a high surface area of up to several hundred m^2 per gram (Warnock et al. 2007), indicating a high percentage of nanopores. A further factor which can influence the pore size or porosity of biochar particles in soil is the adsorption of organic matter on biochar surface which act as a pore blocker (Pignatello et al. 2006). The hyphae of most fungi are 3–6 µm in diameter and it is clear that they can colonize only pores larger than their hyphae diameter (Allen 2007, Ritz 2007, Ottow 2011).

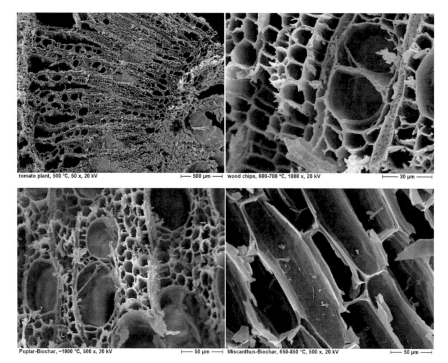

Figure 4. Scanning electron microscopy (SEM) pictures of biochars made of tomato plant, wood chips, poplar and miscanthus produced at different temperature showing the highly diverse porous structure of different feedstocks and temperature, making it an ideal refuge for fungal colonization. Differences regarding the pore size due to the different temperature are not visible (pictures: K. Wiedner 2011).

In addition to the biochar structure itself, elemental and molecular characteristics of biochar surface are a further important criterion for fungal growth on particles. Ascough et al. (2010) investigated the response of two saprophytic white-rot fungi (*Pleurotus pulmonarius* and *Trametes versicolor*) on charcoal as a growth substrate in a short-term experiment (70 days). They found that fungi colonized the surface and to a lesser extent the interior of the charcoal but in different growth forms. The authors concluded that the habitat itself (physical structure) and available nutrients like P, K and Ca on biochar surface were possible reasons for fungal colonization. Saito and Muramota (2002) and Warnock et al. (2007) suggested that depending on the pore size, microorganisms use this refuge to protect themselves against natural predators.

It has been known for many years that biochar can act as a habitat for mycorrhizal fungi (Ogawa and Yamabe 1986, Saito 1990, Gaur and Adholeya 2000, Ezawa et al. 2002). Hockaday et al. (2007) found growth of an unidentified filamentous organism, whose mycelium was of 4 µm in diameter, on 100 years old charcoals. Microbial colonization on naturally produced charcoals from forest wildfires has been found by Pietikäinen et al. (2000), but it is not clear whether it was fungi or bacteria growing on the chars.

Rapid fungal colonization on fresh hydrochars supported an often used paradigm in soil science, e.g., for acid forest soils that soil fungi have a preference for acid substrates. A short-term experiment of fungal growth on hydrochars produced from maize silage by hydrothermal carbonization (HTC) showed a better fungal growth on neutralized hydrochars compared to original acid hydrochars (Fig. 5). Fungi usually grow well at neutral pH

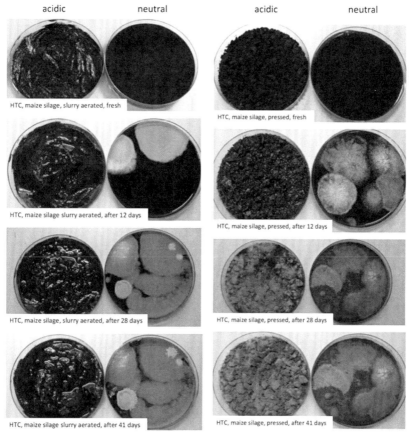

Figure 5. Fungal succession on hydrochars produced from hydrothermal carbonization of maize silage aerated (left) and pressed (right). Please note that the neutralized samples contained more liquid due to neutralization so that the surface looks like an agar plate. The time series show a clear preference of fungal colonization on the neutral hydrochars compared to the acidic ones. Additionally, no fungal colonization occurred within the 41 days on the original aerated hydrochar version (acidic) whereas the acidic pressed hydrochar was colonized by fungi after 28 days. On both neutral hydrochar samples, initial colonization was dominated by *Penicillium* spp. (pictures: K. Wiedner).

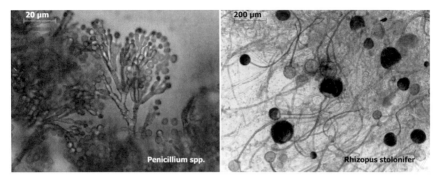

Figure 6. Initial colonization was dominated by species belonging to the *Penicillium* spp. Genera (left). *Rhizopus stolonifer* was determined microscopically on the neutral substrate of the granulated hydrochar (right). Microscopic identification after Domsch et al. 1995 (pictures: K. Wiedner).

but they have better tolerance to acidity than for example bacteria which is very different from preference. It should be mentioned here that hydrochars are initially sterile due to production conditions (180°C and 20 bar for several hours) so that fungal colonization must occur, e.g., by fungal spores from the air or in soil. Interestingly, early colonization was dominated by different *Penicillium* spp. on all samples (Fig. 6), while *Rhizopus stolonifer* was found on the granulated hydrochar sample only. This short-term observation shows clearly the influence of the production conditions and pH value on fungal colonization. However, the question on whether the hydrochars are used by fungi as a carbon source still remains open.

2 Biochar and Pathogenic Fungi Interactions

2.1 Biological Features of Pathogenic Fungi

Pathogenic fungi can be divided into four groups: a) parasitic (growing at the expense of other fungi), b) animal pathogen, c) human pathogen, and d) plant pathogen (or phytopathogen, Fig. 7). The last group may be further splitted into biotrophic and necrotrophic pathogen fungi. Biotrophic fungi need the resource of living host cells, in contrast with the necrotrophic fungi, who kill the infected cells to get energy and grow on dead plant tissue. The penetration into the plant cell involves chemical processes, such as secretion of enzymes (xylanases, cutinases, lipases and cellulases) or mechanical processes through appressoria (infection hyphae). The fungal attack affects plants in different ways. For instance, some phytopathogens reduce the photosynthesis rate, accelerate plant senescence or feed on the plant tissue, like necrotrophic fungi do (Moore et al. 2011). Representatives of necrotrophic fungi are *Botrytis cinerea* (grey mould), *Cochliobolus*

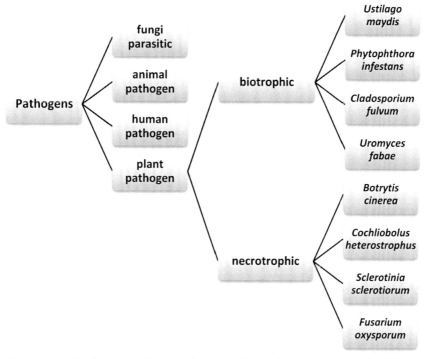

Figure 7. Functional groups of pathogenic fungi. Plant pathogenic fungi are divided into biotrophic and necrotrophic lifestyle. Representative fungal species are given for each group.

heterostrophus (corn leaf blight), *Sclerotinia sclerotiorum* (soft rot) and *Fusarium oxysporum* (vascular wilt). *Ustilago maydis* (maize smut), *Uromyces fabae* (rust), *Cladosporium fulvum* (tomato leaf mould) and *Phytophthora infestans* (potato late blight) are common biotrophic pathogen fungi.

2.2 The Disease Triangle

Healthy plants are able to protect themselves against pathogenic fungi using hypersensitive response. In natural ecosystems, outbreaks of soil-borne diseases and pests are relatively rare compared to agroecosystems (Brussaard 1997). One important reason for plant infestations in agroecosystems is the intensive agriculture dominated by monocultures, which promotes the spreading of soil-borne pathogens. When this is combined with further stress factors, like nutrient shortage (especially nitrogen and phosphorus) or aridity, plants becomes very vulnerable to diseases. If plant roots are infected by a fungal pathogen, the uptake of nutrients is restricted, leading to plant collapse. Besides the infection of plant roots, plant pathogen fungi such as rust, smut fungi, mildew and late blight can infest aerial parts of already weakened plants. Therefore, phytopathogenic fungi are of great importance in agroecosystems because they entail costs up to billions of dollars worldwide every year (Rossman and Palm 2006). Biochar could provide an opportunity to reduce, or even suppress plant disease caused by fungi in agricultural environment. The interaction between environment, host and pathogen is shown by the so-called disease triangle (Fig. 8, Scholthof 2006).

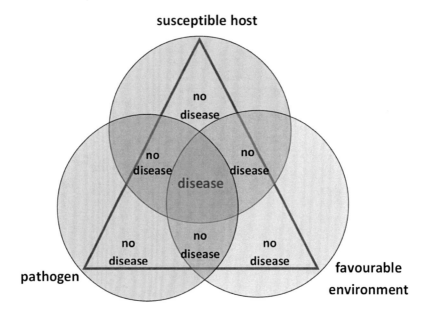

Figure 8. Disease triangle illustrating the three essential factors causing plant diseases. If one of these factors decreases, for example by using biochar, the outbreak of a plant disease is prevented (modified from Moore et al. 2011).

2.3 Biochar Effects on Pathogenic Fungi

It is astonishing that positive effects of charcoal amendments to soil on plant health were already reported 150 years ago (Allen 1846a,b, Retan 1915). Allen (1846a) wrote:

> *'Charcoal as well as lime, often checks rust in wheat, and mildew in other crops; and in all cases mitigates their ravages, where it does not wholly prevent them.'*

The second report of Allen (1846b) describes charcoal as a remedy of potato disease (potato rot) demonstrated in field trials. Retan (1915) reported a reduction of soil-borne phytopathogen diseases of plants after adding charcoal to soil. Interestingly, regarding the effects of charcoal on plant pathogens, recent research starts were it stopped more than 150 years ago.

Among the most problematic pathogenic fungi in agriculture are the species belonging to the genus *Fusarium,* the most abundant mold fungi worldwide. Yield and quality losses, obviation of germination and the release of mycotoxins are only a few problems commonly reported with this filamentous fungus. Matsubara et al. (2002) and Elmer and Pignatello (2011) examined the interaction of biochar with the fungi *Fusarium oxysporum* and *Fusarium proliferatum* on asparagus plants. The authors demonstrated that plants inoculated with mycorrhizal fungi were more resistant against phytopathogens; and even more when biochar was added.

A recent study by Elad et al. (2010) showed a significant decrease of powdery mildew (*Leveillula taurica*) infestation after biochar addition into an organic potting mix and a sandy soil in comparison to no Biochar addition. This leads to the assumption that biochar influences the pathosystem in a different way as other organic matter. Furthermore, the study demonstrated that soil-applied biochar induces systemic resistance to the foliar fungal pathogen *Botrytis cinerea* (gray mold) and *Polyphagotarsonemus latus* (mite pest) on pepper and tomato plants.

The research of the last 30 years has focused on the disease suppressive effects of composted materials. A large body of literature examined the suppression of soil-borne phytopathogen fungi trough composts which are recently reviewed by Noble (2011). The author concluded that the effects of compost amendment to soils on plant pathogens are neutral to beneficial, and that the risk of promoting and introducing diseases appears very small. He also mentioned that a minimum of 15 tonnes compost per hectare is needed to achieve disease suppression and that compost is no panacea against every soil born plant disease. We could therefore expect that a mixing biochar to compost would improve pathogen resistance. This is speculative but conceivable, because biochar and compost may suppress different soil borne pathogens. The disease suppressiveness effects of compost are highly variable. For instance, immature composts release volatile compounds such as sulphur, organic acids or ammonia which are responsible for disease suppression. The suppressive effects of mature composts are predominantly biological, either through microbial antagonism or increased plant host resistance (Noble 2011). It is probable that growth effects of biochar on arbuscular mycorrhizal fungi make the plants more resistant against fungal pathogens (Elmer and Pignatello 2011). Biochar and compost as a mixture could reduce the application of fungicides and pesticides due to its adsorption properties. However, it is known that biochar adsorbs organic molecules such as fungicides and pesticides (Zheng et al. 2010). Consequently, their effectiveness is negatively affected as shown by Andersen (1968), Jordan and Smith (1971), and Yang et al. (2006).

In summary, it has been known for a long time that charcoal addition to soil could increase plant defence against pathogens. The few existing studies showed that biochar has the potential to reduce or prevent the outbreak of fungal plant diseases. The reduction of application of harmful pesticides in agroecosystems through biochar addition seems possible but further research is needed on the adsorption of pesticides by biochar.

3 Biochar and Mycorrhizae Interactions

3.1 The Mycorrhizal Symbiosis

The mutual or parasitic symbiosis between plants (phytobiont) and fungi (mycobiont) are ubiquitous in terrestrial ecosystems and ecologically and economically of great importance (Schüßler 2001). A well-studied form of mutualistic relationship with roots of almost all plant species (90% of land plants) are mycorrhizae (gr. mykes = fungi, rhiza = root). Mycorrhizae improve the growth of plants through mineral and water supply particularly in stress situations. The phytobiont provides carbohydrates, lipids and vitamins for the mycobiont (Ottow 2010). Depending on their morphological and anatomic features, seven classes of mycorrhizae can be distinguished (Fig. 9) (Schwantes 1996, Martin 2001, Smith and Read 2008). Ectomycorrhiza (EM) and endomycorrhiza (mainly arbuscular mycorrhiza, AM) are the most widespread and abundant symbiotic fungi (Smith and Read 2008) and are therefore described in more detail in the following.

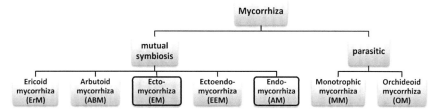

Figure 9. The seven different types of mycorrhizae. Arbuscular mycorrhizal (AM) fungi (outlined in bold) is the symbiosis partner of most arable crops and therefore of high interest in biochar research.

3.2 Ectomycorrhiza (EM)

Main morphological characteristics of EM fungi are the hyphal mantle, covering the short roots including the root tip, and the intercellular Hartig net. The latter is characterized by fungal hyphae growing in between the plant cells between the epidermis and cortical tissue. Fungal hyphae do not penetrate the root more than a few cell layers deep into the cortex (intercellular growing) and do not grow into the host cells. The nutrient exchange between plant and fungi occur through the Hartig net. Polyphosphates, lipids and carbohydrates are stored in the mantle which is more or less thick, depending on the fungal species (Moore et al. 2011). Cell division of plant roots is reduced and the growth of root hair suppressed.

Ectomycorrhizal fungi, belonging to basidio- and ascomycota, are of great importance for forest ecology throughout the world. They provide an essential nutrient supply for conifer and broadleaved trees. In general, the abundance of EM fungi at nutrient-poor locations is much higher than at nutrient-rich locations (Ottow 2011). In short rotation forestry, fast growing tree species like *Salix* (willow) and *Populus* (poplar), have symbiotic associations with EM or AM fungi (Püttsepp et al. 2004, Rooney et al. 2009, Hrynkiewicz et al. 2010).

Research regarding the impact of charcoal on EM fungi was carried out by Japanese scientists since the early 80s and 90s of the last century. In times of increasing need of wood and extensive destroyed forest ecosystems, biochar could be a tool to rehabilitate forests or support measures of reforestation. The application of charcoal powder and a small amount of chemical fertilizer to a young *Pinus thunbergii* stand promoted root growth and mycorrhizal infection (Ogawa and Okimori 2010). It was supposed that the enhanced nutrient availability and water holding capacity were responsible for the increased mycorrhizal formation. In tree nurseries, the use of charcoal as fertilizer to improve soil properties and to promote mycorrhizal growth is well established, at least in Japan (Ogawa 2007).

Makoto et al. (2010) examined the growth response of *Larix gmelinii* seedlings with and without EM inoculation. They reported higher amounts of total and above-ground biomass of seedlings grown in layered charcoal when the trees were in association with EM fungi, together with a higher phosphorus concentration in needles. Solaimann et al. (2010) found an increased mycorrhizal root colonization and beneficial effects on growth and yield of wheat after adding a biochar-mineral fertilizer mixture (composition of the mineral fertilizer is N 8, P 17.6, S 7, Ca 8.6, Cu 0.5 and Zn 0.5%) into a shallow sandy soil, despite periods of drought.

We still do not fully understand how biochar might affect abundance and physiology of mycorrhizal fungi (Warnock et al. 2007). Four mechanisms may be involved: (1) changes of soil nutrient availability, (2) indirect effects on mycorrhiza through the modification of the abundance of other microorganisms in soil, (3) modification of symbiotic signals between phytobiont and mycobiont, (4) provision a habitat for fungi and bacteria (Warnock et al. 2007).

3.3 Endomycorrhiza (AM)

Two-thirds of the plant species have symbiotic associations with fungi of the phylum Glomeromycota (AM fungi), which is also the oldest symbiosis partner of more than 80% vascular plants (Helgason and Fitter 2009, Schüßler 2009). Paleobotanical and molecular sequence data indicate that AM fungi originated during the Ordovician period (~488–433 Ma). Since then, a co-evolution between terrestrial plants and fungi developed (Schüßler and Walker 2011).

Endomycorrhizal fungi are characterized by highly branched intracellular structures (called arbuscules), in contrast to EM fungi where hyphae do not penetrate host cells. These arbuscules increase the contact surface area between plant and fungi and improves nutrient exchange, mainly carbon and phosphate. This is the reason why Arbuscular Mycorrhiza (AM) is often used as a synonym for endomycorrhiza. Another difference to EM fungi is the absence of a fungal mantle and Hartig net and the absence of repression of root hair growth. In addition, all AM fungi are obligately biotrophic their survival is dependent on plants but without host specificity (Moore et al. 2011). In older literature, the term Vesicular-Arbuscular Mycorrhiza (VAM) is used. This is because some genera (e.g., *Glomus, Entrophospora, Acaulospora and Sclerocystis*) form, besides arbuscules, typical storage organs called vesicles. In opposite, representatives of the genus *Gigaspora* and *Scutellospora* only form arbuscules (Ottow 2011). Based on molecular, morphological and ecological characteristics, Arbuscular Mycorrhiza were placed into a new monophyletic phylum, called Glomeromycota (Schüßler 2001).

The high interest in understanding mycorrhizal symbiosis is due to the fact that the beneficial effects of AM fungi on plant performance and soil properties are essential for sustainable management of agricultural ecosystems (Gianinazzi et al. 2010). Therefore, the question of how biochar affects AM symbiosis with arable crops is subject of increasing interest.

Early studies of Japanese researchers showed an increased colonization of crop plants after charcoal application (Saito 1990, Nishio 1996, Saito and Marumoto 2002). In opposite, Warnock et al. (2010) found a negative effect of biochar on AM fungi. The application of three types of biochar (lodgepole pine, peanut shell and mango wood biochars) resulted in a significant decrease of AM fungi abundance from 16.7 to 4.50 m hyphae/cm^{-3} soil (lodgepole), 2.12 to 1.33 m hyphae/cm^{-3} soil (peanut shell) and 19.2 to 4.45 m hyphae/cm^{-3} soil (mango wood), with increasing biochar application rate ranging from 0 to 116.1 tons biochar-C ha^{-1}.

Moreover, Warnock et al. (2007) showed that the most beneficial aspect of AM symbiosis, namely phosphorus-containing nutrient supply, was also affected through biochar addition. The P availability in the lodgepole

Biochar-treated soil significantly declined by about 30% with decreasing AM abundance. In contrast, the P availability in soils with peanut shell and mango wood biochars increased by 101% for the former and by 163% and 208% for the latter biochar, despite decreasing AM fungal abundance. These very important results indicate that biochar application rates, feedstock properties, or production conditions are decisive factors affecting AM fungi abundance and nutrient availability. However, the mechanisms by which these factors changed soil fertility and AM abundance and colonization is still lacking (Warnock et al. 2007).

In addition, it is assumed that biochar acts also as a niche for Mycorrhization Helper Bacteria (MHB) (Warnock et al. 2007). These bacteria trigger morphological and physiological changes of plant roots, facilitating their colonization by mycorrhizal fungi (Rigamonte et al. 2010).

Rillig et al. (2010) tested the effects of application of hydrothermally carbonized (HTC) material in different relative amounts (10 and 20 vol% HTC to soil) on two plant species (*Taraxacum* sp. and *Trifolium repens*) inoculated with *Glomus intraradices*. An application rate of 10% volume of HTC material caused deleterious effects like a declining plant growth, while an addition of 20% volume stimulated the root colonization by the AM fungus. These contradicting results show that more research is needed regarding biochar interactions with AM fungi.

4 Biochar and Saprophytic Fungi Interactions

4.1 The Different Types of Fungal Saprophytes

Saprophytic fungi play a key role in the primary decomposition of plant and animal residues which includes proteins, carbohydrates, fats and lignin. Moreover, they are able to decompose xenobiotic macromolecules in soils and water, including compounds of anthropogenic origin, such as polycyclic aromatic hydrocarbons (PAHs). The secretion of extracellular enzymes (exoenzymes) like ligninases, proteases, cellulases and pectinases enables some fungi to depolymerise and assimilate recalcitrant carbon compounds, such as lignocelluloses, that are found in wood or plant litter. They are degraded by a variety of fungi, *via* three main mechanisms: soft-rot (e.g., *Chaetomium, Ceratocystis*), brown-rot (e.g., *Piptoporus betulinus, Serpula lacrymans*) and white-rot (e.g., *Phanerochaete chrysosporium, Trametes versicolor*). Fungi responsible for white rot are mainly Basidiomycetes and are able to depolymerize lignin or stable aromatic compounds, like PAHs (Field et al. 1992). Key enzymes involved in lignin degradation are lignin peroxidases, manganese peroxidases and laccases (Périé and Gold 1991, Hatakka 1994, Eggert et al. 1997). Decomposition of lignin by basidiomycetes occurs exclusively aerobic, initially co-metabolic (with hemicellulose and cellulose as C- and energy source) and later metabolic and syntroph with fully mineralization and energy production (Ottow 2011). Co-metabolism describes the oxidation of a substance by microorganisms without using it as a source of energy (Leadbetter and Foster 1959).

4.2 Biochar Degradation by Saprophytic Fungi

The chemical structure of biochar (Fig. 1) is comparable to lignin (Fig. 10) or PAH (Fig. 11) comprising (poly)-condensed aromatic moieties and functional groups. Therefore, it is reasonable to assume that the degradation of biochar by saprophytic fungi is similar to the one of complex aromatic molecules, such as lignin or PAHs. In particular, white-rot fungi are able to break down lignin and mineralize highly condensed aromatic structures such as PAHs to smaller molecules that can be assimilated (Cerniglia 1992, Hockaday 2006, Haritash and Kaushik 2009, Chen and Hu 2010). Hence saprophytic fungi could affect stability and longevity of biochar within soil (Jeffrey et al. 2011). Although charcoal-type material can have a residence time in soils of several thousands of years (Glaser and Birk 2012), it can be degraded by saprophytic fungi (Hockaday et al. 2006), who are in addition more abundant in biochar-rich soils such as *terra preta* (Glaser and Birk 2012). Specific research regarding biochar degradation with selected saprophytic fungi is rare. However, previous studies showed that wood-decaying fungi are capable to degrade low-rank coal (Cohen and Peter 1982, Fakoussa and Hofrichter 1999, Hofrichter et al. 1999). Several mechanisms, like the production of oxidative enzymes (peroxidases, laccases), hydrolytic enzymes (esterases), alkaline metabolites (e.g., ammonia and biogene amines) and natural chelators (e.g., ammonium oxalate and siderophores) are involved in this degradation (Fakoussa and Hofrichter 1999). Nevertheless, up to now little is known if any of these chemical or biological mechanisms are involved in biochar degradation (Hockaday 2006). It can be assumed that biodegradation of biochar occurs with similar mechanisms as PAH degradation. This is based on the fact that PAHs and biochar arise under common synthetic processes (pyrolysis) and the chemical structure is similar, especially the polyaromatic backbone (Figs. 1 and 11). In contrast to microbial biochar degradation, mechanisms of PAH biodegradation are much better understood (e.g., Johnsen et al. 2005, Haritash and Kaushik 2009). Because of its highly recalcitrant chemical structure, biochar is not expected to be a direct carbon source for microorganisms (Birk et al. 2009). Therefore, it is suggested that highly condensed aromatic structures like PAHs (Fig. 11) or biochar (Fig. 1) are degraded in a co-metabolic process (Fig. 11). However, Ascough et al. (2010) showed that the breakdown of the structure of charcoal itself does not supply a readily available source of nutrients to the species *Pleurotus pulmonarius* and *Trametes versicolor*. In contrast, Wengel et al. (2006) found that the basidiomycete saprophyte *Schizophyllum commune* oxidized biochars (mixture of beeches and oaks, produced at 400°C) with a highly condensed aromatic structure through the secretion of exoenzymes like peroxidases and laccases. Additionally, diversity and function of saprophytic soil fungi are sensitive to disturbance, pollution and environmental change (Deacon et al. 2006). Therefore, biochar amendment could lead to unexpected behaviour regarding to their ability of biodegradation. These two contrasting studies demonstrate the importance of further investigations regarding biochar stability in soil. Biochar has the ability to promote fungal abundance in soil, as a study of Steinbeiss et al. (2009) showed after the addition of ^{13}C-labeled yeast (5% N) and glucose (0% N) as feedstock material for biochar into arable and forest soils. The isotopic shifts of PLFA biomarkers showed that there was a strong yeast-derived biochar uptake by fungi in both soils of 14.8 and 9.5‰, respectively. Fungi were less involved in the decomposition of glucose-derived biochar in both soils (isotopic shift for arable soil = 1.6‰ and for forest soil 4.8‰).

There is disagreement whether besides the common saprophytic fungi like soft, brown and white rot fungi, some species of mycorrhizal fungi, like ericoid mycorrhiza, ectomycorrhiza (EM) and arbuscular mycorrhiza (AM), are also able to decompose organic matter in soil (Talbot et al. 2008). Abiotic stress such as extreme temperatures, heavy metal toxicity or oxidative stress (Singh et al. 2011) leads to low photosynthate supply from the host for the mycorrhizal fungi (Baldrian 2009). Due to the increase of the enzymatic activity of the mycorrhizal fungi during these periods, some authors suggest that the fungi supply the carbon from soil. Baldrian (2009) mentions two reasons against the decomposition of soil carbon by EM fungi: (i) EM fungi appear in deeper soil horizons in which carbon has low energetic and (ii) the production of ligninolytic enzymes and cellulases by EM fungi is much weaker compared to basidiomycetes. In contrast, recent studies (e.g., Hodge et al. 2001, Tu et al. 2006) suggest that AM fungi are able to degrade complex organic materials like grass leaves. Due the contradictory meanings and the lack of studies regarding a possible potential of mycorrhizal fungi to decompose stable organic compounds in soil, it is important to investigate the behaviour of mycorrhizal fungi during abiotic stress situations regarding their possible ability on biochar degradation.

In summary, due to the lack of available data about saprophytic decomposition of biochar it is necessary and helpful to have a closer look at related fields of research to find new approaches. Even the above-mentioned examples showed that soil and biochar properties play an important role on biochar degradation by soil fungi. Fungal abundance and diversity which is affected through biochar addition is also of importance. Although there is clear evidence that biochar can be stable in soils over thousands of years (2,000 years of mean residence time and 1,400 years of half-life time for ryegrass biochar, see Kuzyakov et al. 2009, Glaser and Birk 2012), there is also evidence that biochar can be decomposed rapidly by fungi (e.g., Wengel et al. 2006). Nevertheless, details about the degradation process of biochar are far from being understood.

4.3 Enzymes Involved in Degradation of Complex Organic Molecules

Being a component of vascular plants including ferns, lignin is the most abundant renewable aromatic material on earth (Kirk and Farrell 1987). It is an aromatic polymer with a complex structure made up of phenyl-propane-based monomers linked via a variety of bonds that bind cell-wall components together (Kirk and Farrell 1987, Freudenberg and Neish 1969, Fig. 10). Many authors suggest that the high recalcitrance of lignin against microbial decomposition is due to several persistence factors, such as high amount of benzene rings and a high percentage (50 to 70%) of stable α- and β-O-4-aryl ether rings (R-O-R) and biphenyl bonds (e.g., Ottow 2011). However, recent studies proved that lignin may turn over within years to decades in arable soil (Dignac et al. 2005, Glaser 2005, Hoffmann et al. 2009). Lignin forms a protective shield around cellulose and protects other polymers from attack. Therefore, a delignification step is

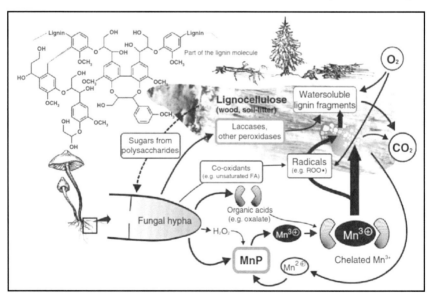

Figure 10. Chemical structure of lignin and biochemical reactions of fungal lignin degradation (Hatakka and Hammel 2010, with permission).

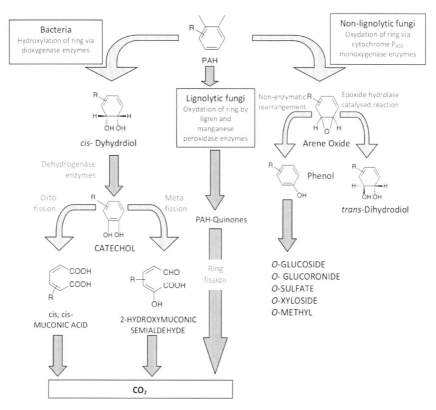

Figure 11. Pathways for the catabolism of PAHs by bacteria, lignolytic fungi and non-lignolytic fungi with bacteria (modified from Cerniglia 1992).

necessary for fungi to access cellulose. White-rot fungi (Basidiomycota and Ascomycota) like *Fomes fomentarius*, *Phellinus igniarius* or *Trametes versicolor* are able to perform this enzymatic process. Most fungi decompose lignin and cellulose simultaneously (e.g., *Phellinus pini*). Species like *Xylobolus frustulatus* or *Heterobasidiom annosum* initially degrade lignin, which implicates relative enrichment of the substrate in cellulose (selective white-rot). Every mechanism of degradation occurs under aerobic and energy intensive conditions (co-metabolism). Hence, lignin degradation is always a co-metabolic process combined with easily available carbon sources. The depolymerisation of lignin requires a panel of enzymes including (i) lignin peroxidases, (ii) manganese peroxidases and (iii) laccases (Moore et al. 2011, Fig. 10). Depending on the production of these three types of enzymes in combination, white-rot fungi are divided into three groups: (i) lignin-manganese peroxidase group (e.g., *Phanerochaete chrysosporium*), (ii) manganese peroxidase-laccase group (e.g., *Dichomitus squalens*), and (iii) lignin peroxidase group (*Junghuhnia separabilima*) (Hatakka 1994).

Polycyclic Aromatic Hydrocarbons (PAHs) is a group of several hundred single organic substances which are ubiquitous in the environment. The structure of PAH molecules consists of two or more condensed benzene rings. Microbial decay (biodegradation) of PAH into less complex metabolites and complete degradation are the major degradation processes besides volatilization, photolysis and chemical degradation (Haritash and Kaushik 2009). Extracellular enzymes responsible for PAH degradation are the same as for lignin degradation: laccase, lignin peroxidase and manganese peroxidase (Haritash and Kaushik 2009, Figs. 10 and 11). The low substrate specificity of these enzymes makes them suitable to degrade different organic compounds (Haritash and Kaushik 2009). Therefore it is obvious to speculate that saprophytic soil fungi have a great potential to degrade stable carbon compounds like biochar.

Some non-ligninolytic fungi, like *Aspergillus niger, Cunninghamella elegans* and *Chrysosporium pannorum*, are also able to produce cytochrome P_{450} monooxygenase enzymes that oxidise PAHs (Bamforth and Singleton 2005). In addition, some species of soil bacteria (for instance some belonging to the genus *Pseudomonas*, gram-negative bacteria) are also able to degrade aromatic hydrocarbons using dioxygenase enzymes (Mrozik et al. 2003). Moreover, in contrast to fungi, bacteria can use PAHs metabolites as only carbon source. It is suggested that bacteria and fungi could have a co-metabolic cooperation (Fig. 11) for PAH degradation (Johnsen et al. 2005). The mechanisms of PAH degradation by bacteria, ligninolytic and non-ligninolityc fungi differ fundamentally (Fig. 11). Nevertheless, all three mechanisms are based on the oxidation of the aromatic ring and the systematic breakdown of the compound to PAH metabolites (Bamforth and Singleton 2005, Fig. 11). The co-metabolic degradation process by fungi is currently insufficiently clarified and it is therefore difficult to estimate if bacteria or fungi are more involved in PAH decay and therefore also in a possible biochar degradation. It can be assumed that the mechanisms of biochar degradation are very similar to the degradation mechanisms of PAHs, as outlined in Fig. 11.

4.4 The Role of Biochar in Co-metabolism and Priming Effect

Concerning the rising atmospheric CO_2 concentration, it is of scientific interest to understand whether soils act as net carbon sink or source in the global carbon cycle (Lal 2004, Le Quéré et al. 2009). Some studies suggest that biochar has the potential to sequester C in soil ecosystems and can act as long-term C sink due to its high stability (e.g., Glaser et al. 2002, Lehmann 2007, Zavalloni et al. 2011). However, several complex and poorly understood mechanisms in soils -like co-metabolism and priming effect—may influence labile C compounds and the stability of biochar. The term co-metabolism

(Jensen 1963) is based on the term 'co-oxidation' and was firstly coined by Leadbetter and Foster (1959), describing the oxidation of a substance by microorganisms without utilization of the derived energy. Normally, this occurs only in parallel with the metabolism of substances of which the organism has an immediate energy supply. In the early 20th century, enhanced soil organic carbon (SOC) mineralization after the input of fresh organic matter into soil was observed by Löhnis (1926). This phenomenon was quantified by Bingeman et al. (1953) as 'priming effect'. Up to now, despite the large body of literature, the complex mechanisms responsible for priming effect are only partly understood. One possible involved mechanism is co-metabolism (Kuzyakov et al. 2000). Adding fresh organic matter into soil stimulates the growth of microorganisms and therewith the presence of unspecific enzymes. For instance, a simple litterbag experiment of Wardle et al. (2008) in the boreal forest in northern Sweden indicated that the presence of charcoal can enhance the decomposition of soil organic matter within a decade. It is still unclear which microorganisms prime soil organic matter (SOM) decomposition (fungi, bacteria or other microbes) and how far they interact with each other (Kuzyakov 2010). There are several theories which try to explain their strategy and interaction for instance the "r/K selection" or "life trait history" theories. A final explanation about which microorganisms prime SOM decomposition still remains open.

Numerous experiments on priming effects were conducted in recent years and were reviewed by Kuzyakov (2010). He summarized that the mechanisms, microbial groups, hot spots and time periods regarding the phenomena of priming effect is only partly answered in the studies. One of these studies which is interesting with respect to biochar is an *in vitro* experiment by Hamer et al. (2004), who added glucose and a glucose/nutrient $[(NH_4)_2SO_4 + KH_2PO_4]$ mixture to different charred materials mixed with sand. The biochar mineralization was strongly enhanced after the two glucose additions (between 0.6 and 1.2%) compared to the mineralization in the controls without glucose (0.3 to 0.8%). One explanation of these results could be an increase of the amount of bacteria and fungi inside and outside biochar particles (Chenu et al. 2001). Despite these results, it has to be considered that added organic matter in agroecosystems like manures or crop residues are much more recalcitrant and chemically complex than glucose (Kuzyakov et al. 2009).

Nevertheless, the knowledge about soil priming effect after biochar addition is still very limited. Some recent studies investigated the effect of labile organic matter addition to soil on biochar mineralization. Few studies examined the potential priming effect of biochar on native organic matter in soils. The results are contradictory. For instance, studies of Major et al. (2010) and Novak et al. (2010) showed no or a small priming effect after labile organic matter (LOM) addition to biochar. Keith et al. (2011) performed an *in vitro* short-term experiment (120 days 20°C) with different mixture rates of biochar and sugar cane residues as LOM source in a smectitic clayey soil. For the evaluation of different mineralization rates, the $\delta^{13}C$ values of biochars were distinct from the ones of labile organic matter and soil. The results showed that the total mineralization rate of the 450°C biochar within each LOM treatment was 1.5 to 1.9 times higher compared with the total mineralization rate of 550°C biochar. The mineralization rates of biochar were positively correlated with application rates of labile organic matter. In turn, labile organic matter mineralization rate decreased with biochar application. Nevertheless, the interactive priming of biochar and labile organic matter stabilized at later stages (Keith et al. 2011). Another study of Zavalloni et al. (2011) examined the mineralization of hardwood biochar mixed with wheat straw in soil (soil with 5% biochar and 0.5% wheat straw). Wheat straw did not induce priming effect on biochar C decomposition, which was measured as total net CO_2 emission by soil respiration. The potential of biochar to induce priming effect on native organic carbon was examined by Cross and Sohi (2011) and Luo et al. (2011). Cross and Sohi (2011) incubated C4 biochars produced at different conditions mixed with three C3 soils with different organic matter contents. CO_2 flux rates and the corresponding isotopic attribution showed that labile Biochar compounds were mineralized but no priming of native organic matter occurred. A few soils showed a negative priming effect after biochar addition compared with soils without biochar. The authors suggested a stabilization of labile soil carbon in these soils. Luo et al. (2011) showed that biochar could have a positive priming effect on native organic C, by using biochars made of *Miscanthus giganteus* and ryegrass and produced at 350 and 700°C. Different biochar mixtures with and without ryegrass were added to clay-loam soils with a pH value of 3.7 and 7.6. After 87 days of biochar (350°C) incubation, priming effects equivalent to 250 and 319 μg CO_2-C^{-1} were caused in both soils. The highest priming effects appeared immediately after biochar addition and were lower for biochar produced at 700°C. The addition of ryegrass primed the mineralization of both biochar types significantly. The enhanced but not long lasting mineralization rate of biochar directly after adding labile organic matter could be explained by co-metabolism, which is a short-term mechanism. Zimmerman et al. (2011) incubated five soils from Florida with different biochar types (Table 1). The results ranged from a negative to a positive priming effect (Table 1). Authors suggested that the reason for the negative priming effect was a limitation of microbial activity because of the adsorption of organic nutrients and organic matter on biochar surface. A further experiment with ^{14}C-labeled ryegrass residues (*Lolium perenne*) biochar was examined by Kuzyakov et al. (2009). They added biochar to the Ah layer of a Haplic Luvisol and added glucose. The decomposition rate of biochar in both substrates increased up to six times

Table 1. Soil and biochar (BC) properties are important factors regarding a possible positive or negative priming effect triggered by biochar addition to soil. Additionally, results of studies are difficult to compare due to different experimental setups, measuring methods or periods. Nevertheless, the low rates of biochar loss show that biochar is stable in soil and therefore suitable for long-term C sequestration.

Experiment	Soil type	pH value soil	BC material	pH value BC	LOM type	BC and LOM loss	Reference
BC decomposition rates based on $^{14}CO_2$ sampled 44 times during the 3.2 years	Haplic Luvisol	6	perennial ryegrass (*Lolium perenne*), 400 °C		glucose	< 4.5 % BC ↑	(Kuzyakov et al. 2009)
Fate of BC on arable soil surface at 26 °C annual mean temperature over two years, measuring CO_2 (δ13C)	Typic Haplustox		mango tree wood (*Mangifera indica* L.), 400-600 °C	8.74, 8.92		< 3 % BC ↑↓	(Major et al. 2010)
BC mineralization in a sandy soil within 25 and 67days measuring CO_2 fluxes	Norfolk loamy sand	4.8	pecan-shell, 700 °C	7.5		0.18 to 19.91 µmol CO_2 m^{-2} s^{-1} ↑	(Novak et al. 2010)
C mineralization from biochar and LOM/soil using contrasting δ^{13}C values after 120 days at 20 °C	smectite-rich clayey soil	8.1	wood, 450, 550 °C	8.6, 9.9	sugar cane	0.4-1.1 % BC, 4.2-13.6 % LOM/soil ↑	(Keith et al. 2011)
BC mineralization and wheat straw mixtures in soil after 84 days	Chromi-Endo-skeletic Cambisol	7.2	hardwood, 500 °C	7.2	wheat straw	2.8 % BC, 56 % LOM ↑	(Zavalloni et al. 2011)

Table 1. contd....

Table 1. contd....

Experiment	Soil type	pH value soil	BC material	pH value BC	LOM type	BC and LOM loss	Reference
CO_2 flux rates and $\delta 13C$ of a BC set in soils with different OM status after 14 days at 30 °C	C3 soils; fallow, arable and grassland soils		sugarcane bagasse and 1 sugarcane trash (C4), 350, 450, 550 °C			0.24-4.58 % BC ↑↓	(Cross and Sohi 2011)
Native OM mineralization after BC addition and the effect of ryegrass addition on BC mineralization after 180 days	clay-loam soil	3.7 , 7.6	*Miscanthus giganteus* (C4), 350, 700 °C	7.81, 10.8	ryegrass	0.14% and 0.18% BC 700 °C; 0.61% and 0.84% BC 350 °C ↑	(Luo et al. 2011)
BC mineralization in five different soils measured by CO_2 ($\delta^{13}C$) evoultion within 505 days	two Alfi sols, two Entisols, Mollisol	4.7, 3.9, 5.3, 5.1, 7.6	oak (*Quercus laurifolia*), pine (*Pinus taeda*) and bubinga (*Guibourtia demeusei*), Eastern gamma grass (*Tripsacum dactyloides*), bagasse (sugar cane), 250, 400, 525, 600 °C			1.4 % BC per year ↑↓	(Zimmerman et al. 2011)

↑postive priming effect; ↓ negative priming effect

after adding glucose but declined rapidly, and thus did not influence longer term biochar mineralization (Table 1). The above-mentioned studies and the partly contradictory results show that we still need more research regarding the basic mechanisms of co-metabolism and priming effects and the interactive priming potential of biochar. The nature of interaction between native soil organic matter and labile organic matter with biochar obviously differs. For instance, soils with low native organic matter contents show priming after biochar addition (Keith et al. 2011). Besides this, feedstock type, production temperature, chemical and physical properties of biochars are important factors influencing priming effect, as much as soil type and their associated characteristics can be (e.g., pH, soil moisture, texture, temperature; Table 1). Moreover, it is proved that the addition of biochar stimulates the microbial activity in soil (Steiner et al. 2004). Although it is still unclear which microorganisms prime the degradation of soil organic matter (Kuzyakov 2010), it is certain that fungi are substantially involved (Fontaine et al. 2003).

Despite the positive priming effect noticed in some studies after biochar addition on soil organic C, there is evidence that biochar increases the total soil organic C content and the loss is compensated (Luo et al. 2011).

4.5 Fungal Potential of Biochar Production

Besides biochar degradation by soil fungi, there is scientific evidence for fungal production of biochar or at least compounds which are chemically identical to biochar (Glaser and Knorr 2008). Aspergillin, the black pigment of *Aspergillus niger* being ubiquitous in soils contains condensed aromatic structures (Lund et al. 1953) similar to those of biochar (Brodowski et al. 2005). The contribution of aspergillin to soil biochar inventory was estimated to be negligible in a study by Brodowski et al. (2005). However, Glaser and Knorr (2008) showed by ^{13}C labelling and incubation in the absence of increased temperature ranging from 1 month to 23 years followed by compound-specific ^{13}C analyses of biochar that up to 25% in total (after 23 years) or up to 9% annually of biochar in soils are formed biologically. Further proof for the biological origin of biochar was correlation of newly synthesized biochar (containing the ^{13}C label) with incubation time and temperature regime (temperate or tropical), both confirming that biochar is produced *in situ* in soils. Although there is unambiguous proof that *Aspergillus niger* can produce biochar, nothing is known on the potential of other soil fungi (and bacteria) for biological biochar production.

Acknowledgements

This work was partly funded by the EuroChar project of the European Community (FP7-ENV-2010 Project ID 265179) and the ClimaCarbo project of the German Ministry of Education and Research (01LY1110B). We kindly acknowledge Arne Stark and Robert Maas for providing hydrochar samples for fungal growth tests.

References

Allen, M.F. 2007. Mycorrhizal fungi: Highways for water and nutrients in arid soils. Vad. Zone J. 6: 291.

Allen, R.L. 1846a. American agriculturist. Saxton & Miles, New York, USA.

Allen, R.L. 1846b. A brief compendium of American agriculture. Saxton & Miles, New York. USA.

Andersen, A.H. 1968. The inactivation of simazine and linuron in soil by charcoal. Weed Res. 8: 58–60.

Ascough, P.L., C.J. Sturrock, and M.I. Bird. 2010. Investigation of growth responses in saprophytic fungi to charred biomass. Isotopes Env. Health Studies 46: 64–77.

Baldrian, P. 2009. Ectomycorrhizal fungi and their enzymes in soils: is there enough evidence for their role as facultative soil saprotrophs? Oecologia 161: 657–660.

Bardgett, R., W. Bowman, R. Kaufmann, and S. Schmidt. 2005. A temporal approach to linking aboveground and belowground ecology. Trends Ecol. & Evol. 20: 634–641.

Bamforth, S.M. and I. Singleton. 2005. Bioremediation of polycyclic aromatic hydrocarbons: current knowledge and future directions. J. Chem. Technol. Biotechnol. 80: 723–736.

Bingemann, C.W., J.E. Varner, and W.P. Martin. 1953. The effect of the addition of organic materials on the decomposition of an organic soil. Soil Sci. Soc. Am. Proc. 17: 34–38.

Birk, J.J., C. Steiner, W.C. Teixeira, W. Zech, and B. Glaser. pp. 309–324. Microbial response to charcoal amendments and fertilization of a highly weathered tropical soil. *In*: Woods, W.I., W.G. Teixeira, J. Lehmann, C. Steiner, A. WinklerPrins, and L. Rebellato [eds.]. 2009. Amazonian Dark Earths: Wim Sombroek' s Vision. Springer, Berlin, Germany.

Brodowski, S., W. Amelung, L. Haumaier, C. Abetz, and W. Zech. 2005. Morphological and chemical properties of black carbon in physical soil fractions as revealed by scanning electron microscopy and energy-dispersive X-ray spectroscopy. Geoderma 128: 116–129.

Brussaard, L. 1997. Biodiversity and ecosystem functioning in soil. Ambio 26: 563–570.

Cerniglia, C.E. 1992. Biodegradation of polycyclic aromatic hydrocarbons. Biodegradation 3: 351–368.

Chen, B.W. and D. Hu. 2010. Biosorption and biodegradation of polycyclic aromatic hydrocarbons in aqueous solutions by a consortium of white-rot fungi. J. Haz. Mat. 179: 845–851.

Chenu, C., J. Hassink, and J. Bloem. 2001. Short-term changes in the spatial distribution of microorganisms in soil aggregates as affected by glucose addition. Biol. Fert. Soils 34: 349–356.

Cohen, M. and G.D. Peter. 1982. Degradation of coal by the fungi *Polyporus versicolor* and *Poria monticola*. Appl. Env. Microbiol. 44: 23–27.

Cross, A. and S.P. Sohi. 2011. The priming potential of biochar products in relation to labile carbon contents and soil organic matter status. Soil Biol. Biochem. 43: 2127–2134.

Czimczik, C.I., C.M. Preston, M.W.I. Schmidt, R.A. Werner, and E.-D. Schulze. 2002. Effects of charring on mass, organic carbon, and stable carbon isotope composition of wood. Org. Geochem. 33: 1207–1223.

Deacon, L.J., E. Janie Pryce-Miller, J.C. Frankland, B.W. Bainbridge, P.D. Moore, and C.H. Robinson. 2006. Diversity and function of decomposer fungi from a grassland soil. Soil Biol. Biochem. 38: 7–20.

Dighton, J. 2003. Fungi in Ecosystem Processes. CRC Press, New Jersey, USA.

Dignac, M.-F, H. Bahri, C. Rumpel, D.P. Rasse, G. Bardoux, and J. Balesdent. 2005. Carbon-13 natural abundance as a tool to study the dynamics of lignin monomers in soil: an appraisal at the Closeaux experimental field (France). Mechanisms and regulation of organic matter stabilisation in soils. Geoderma 128: 3–17.

Domsch, K.H., W. Gams, and T.-H. Anderson. 1995. Compendium of soil fungi. Academic Press. London.

Downie, A., A. Crosky, and P. Munroe. Physical properties of biochar. pp. 13–32. *In:* J. Lehmann and S. Joseph [eds.]. 2009. Biochar for Environmental Management: Science and Technology. Earthscan, London, UK.

Eggert, C., U. Temp, and K.-E. L. Eriksson. 1997. Laccase is essential for lignin degradation by the white-rot fungus *Pycnoporus cinnabarinus*. FEBS Letters 407: 89–92.

Elad, Y., D.R. David, Y.M. Harel, M. Borenshtein, H.B. Kalifa, A. Silber, and E.R. Graber. 2010. Induction of systemic resistance in plants by biochar, a soil-applied carbon sequestering Agent. Phytopath. 100: 913–921.

Elmer, W.H. and J.J. Pignatello. 2011. Effect of biochar amendments on mycorrhizal associations and fusarium crown and root rot of asparagus in replant soils. Plant Disease 95: 960–966.

Ezawa, T., K. Yamamoto, and S. Yoshida. 2002. Enhancement of the effectiveness of indigenous arbuscular mycorrhizal fungi by inorganic soil amendments. Soil Sci. Plant Nut. 48: 897–900.

Fakoussa, R.M. and M. Hofrichter. 1999. Biotechnology and microbiology of coal degradation. Appl. Microbiol. Biotechnol. 52: 25–40.

Field, J.A., E. de Jong, G. Feijoo Costa, and J.A. de Bont. 1992. Biodegradation of polycyclic aromatic hydrocarbons by new isolates of white rot fungi. Appl. Environ. Microbiol. 7: 2219–2226.

Fontaine, S., A. Mariotti, and L. Abbadie. 2003. The priming effect of organic matter: a question of microbial competition? Soil Biol. and Biochem. 35: 837–843.

Freudenberg, K. and A.C. Neish. 1969. Constitution and Biosynthesis of Lignin. Springer, New York, USA.

Gaur, A. and A. Adholeya. 2000. Effects of the particle size of soil-loess substrates upon AM fungus inoculum production. Mycorrhiza 10: 43–48.

Gianinazzi, S., A. Gollotte, M.-N. Binet, D. Tuinen, D. Redecker, and D. Wipf. 2010. Agroecology: the key role of arbuscular mycorrhizas in ecosystem services. Mycorrhiza 20: 519–530.

Giri, B., P.H. Giang, R. Kumari, R. Prasad, A. Varma. 2005. Microbial Diversity in Soils. *In:* Buscot, F. and A. Varma. Microorganisms in Soils: Roles in Genesis and Functions. Soil Biology, Volume 3. Sringer, Heidelberg.

Glaser, B. 2005. Compound-specific stable-isotope (delta C-13) analysis in soil science. J. Plant Nutr. Soil Sci. 168: 633–648.

Glaser, B., J. Lehmann, and W. Zech. 2002. Ameliorating physical and chemical properties of highly weathered soils in the tropics with charcoal—a review. Biol. Fert. Soils 35: 219–230.

Glaser, B. and K.H. Knorr. 2008. Isotopic evidence for condensed aromatics from non-pyrogenic sources in soils—implications for current methods for quantifying soil black carbon. Rapid Commun. Mass Spectrom. 22: 935–942.

Glaser, B. and J.J. Birk. 2012. State of the scientific knowledge on properties and genesis of Anthropogenic Dark Earths in Central Amazonia (terra preta de Índio). Geochim. Cosmochim. Acta 82: 39–51.

Grossman, J.M., B.E. O' Neill, S.M. Tsai, B. Liang, E. Neves, J. Lehmann, and J.E. Thies. 2010. Amazonian anthrosols support similar microbial communities that differ distinctly from those extant in adjacent, unmodified soils of the same mineralogy. Microb. Ecol. 60: 192–205.

Hallett, P.D., L. Lichner, A. Cerdà, W. Zhang, J. Niu, and V.L. Morales. 2010. Transport and retention of biochar particles in porous media: effect of pH, ionic strength, and particle size. Ecohydrol. 3: 497–508.

Hamer, U., B. Marschner, S. Brodowski, and W. Amelung. 2004. Interactive priming of black carbon and glucose mineralization. Organ. Geochem. 35: 823–830.

Haritash, A.K. and C.P. Kaushik. 2009. Biodegradation aspects of polycyclic aromatic hydrocarbons (PAHs): A review. J. Hazard. Mat. 169: 1–15.

Hatakka, A. 1994. Lignin-modifying enzymes from selected white-rot fungi: production and role from in lignin degradation. FEMS Microbiol. Rev. 13: 125–135.

Hatakka, A. and K. Hammel. 2010. Fungal biodegradation of lignocelluloses. *In:* The MYCOTA Vol. X, Springer, Berlin, Heidelberg.

Hawksworth, D.L. 2001. The magnitude of fungal diversity: the 1.5 million species estimate revisited. Myc. Res. 105: 1422–1432.

Helgason, T. and A.H. Fitter. 2009. Natural selection and the evolutionary ecology of the arbuscular mycorrhizal fungi (Phylum Glomeromycota). J. Exp. Bot. 60: 2465–2480.

Hockaday, W.C. 2006. The organic geochemistry of charcoal black carbon in the soils of the University of Michigan Biological Station. PhD thesis, Ohio State University, USA.

Hockaday, W.C., A.M. Grannas, S. Kim, and P.G. Hatcher. 2006. Direct molecular evidence for the degradation and mobility of black carbon in soils from ultrahigh-resolution mass spectral analysis of dissolved organic matter from a fire-impacted forest soil. Org. Geochem. 37: 501–510.

Hockaday, W.C., A.M. Grannas, S. Kim, and P.G. Hatcher. 2007. The transformation and mobility of charcoal in a fire-impacted watershed. Geochim. Cosmochim. Acta 71: 3432–3445.

Hodge, A., C.D. Campbell, and A.H. Fitter. 2001. An arbuscular mycorrhizal fungus accelerates decomposition and acquires nitrogen directly from organic material. Nature 413: 297–299.

Hofmann, A., A. Heim, B.T. Christensen, A. Miltner, M. Gehre, and M.W.I. Schmidt. 2009. Lignin dynamics in two ^{13}C-labeled arable soils during 18 years. Europ. J. Soil Sci. 60: 250–257.

Hofrichter, M., D. Ziegenhagen, S. Sorge, R. Ullrich, F. Bublitz, and W. Fritsche. 1999. Degradation of lignite (low-rank coal) by ligninolytic basidiomycetes and their manganese peroxidase system. Appl. Microbiol. Biotechnol. 52: 78-84.

Hoormann, J.J. 2011. The Role of Soil Fungus. The Ohio State University.

Hrynkiewicz, K., C. Baum, P. Leinweber, M. Weih, and I. Dimitriou. 2010. The significance of rotation periods for mycorrhiza formation in short rotation coppice. Forest Ecol. Management 260: 1943–1949.

Jeffery, S., F.G.A. Verheijen, M. Van der Velde, and A.C. Bastos. 2011. A quantitative review of the effects of biochar application to soils on crop productivity using meta-analysis. Agriculture Ecosystems & Environment 144: 174–187.

Jensen, H.L. 1963. Carbon nutrition of some microorganisms decomposing halogen-substituted aliphatic acids. Acta Agr. Scand. 13: 404–412.

Johnsen, A.R., L.Y. Wick, and H. Harms. 2005. Principles of microbial PAH-degradation in soil. Env.Pollution 133: 71–84.

Jordan, P.D. and L.W. Smith. 1971. Adsorption and deactivation of atrazine and diuron by charcoals. Weed Sci. 19: 541–544.

Keith, A., B. Singh, and B.P. Singh. 2011. Interactive priming of biochar and labile organic matter mineralization in a smectite-rich soil. Environ. Sci. Technol. 45: 9611–9618.

Kim, J.-S., S. Sparovek, R.M. Longo, W.J. De Melo, and D. Crowley. 2007. Bacterial diversity of terra preta and pristine forest soil from the Western Amazon. Soil Biol. Biochem. 39: 648–690.

Kirk, T.K. and R.L. Farrell. 1987. Enzymatic "combustion": The microbial degradation of lignin. Annu. Rev. Microb. 41: 465–501.

Kuzyakov, Y., J.K. Friedel, and K. Stahr. 2000. Review of mechanisms and quantification of priming effects. Soil Biol. Biochem. 32: 1485–1498.

Kuzyakov, Y., I. Subbotina, H. Chen, I. Bogomolova, and X. Xu. 2009. Black carbon decomposition and incorporation into soil microbial biomass estimated by ^{14}C labeling. Soil Biol. Biochem. 41: 210–219.

Kuzyakov, Y. 2010. Priming effects: Interactions between living and dead organic matter. Soil Biol. Biochem. 42: 1363–1371.

Lal, R. 2004. Soil Carbon Sequestration Impacts on Global Climate Change and Food Security. Science 304: 1623–1627.

Leadbetter, E.R. and J.W. Foster. 1959. Oxidation products formed from gaseous alkanes by the bacterium *Pseudomonas methanica*. Arch. Biochem. Biophys. 82: 491–492.

Lehmann, J. 2007. A handful of carbon. Nature 447: 143–144.

Lehmann, J. and S. Joseph. 2009. Biochar for Environmental Management, Science and Technology. Earthscan, London. UK.

Le Quéré, C., M.R. Raupach, J.G. Canadell, and G. Marland. 2009. Trends in the sources and sinks of carbon dioxide. Nature Geosci. 2: 831–836.

Löhnis, F. 1926. Nitrogen availability of green manures. Soil Sci. 22: 253–290.

Lund, N.A., A. Robertson, and W.B. Whalley. 1953. The chemistry of fungi. Part XXI. Asperxanthone and a preliminary examination of aspergillin. J. Chem. Soc. 1953: 2434–2439.

Luo, Y., M. Durenkamp, M. de Nobili, Q. Lin, and P.C. Brookes. 2011. Short-term soil priming effects and the mineralization of biochar following its incorporation to soils of different pH. Soil Biol. Biochem. 43: 2304–2314.

Martin, F. 2001. Frontiers in molecular mycorrhizal research—genes, loci, dots and spins. New Phytol. 150: 499–505.

Major, J., J. Lehmann, M. Rondon, and C. Goodale. 2010. Fate of soil-applied black carbon: downward migration, leaching and soil respiration. Global Change Biol. 16: 1366–1379.

Makoto, K., Y. Tamai, Y.S. Kim, and T. Koike. 2010. Buried charcoal layer and ectomycorrhizae cooperatively promote the growth of *Larix gmelinii* seedlings. Plant Soil 327: 143–152.

Matsubara, Y., N. Hasegawa, and H. Fukui. 2002. Incidence of Fusarium root rot in asparagus seedlings infected with arbuscular mycorrhizal fungus as affected by several soil amendments. J. Jap. Soc. Horticult. Sci. 71: 370–374.

Moore, D., G.D. Robson, and A.P.J. Trinci. 2011. 21st Century Guidebook to Fungi. Cambridge, New York, USA.

Mrozik, A., Z. Piotrowska-Seget, and S. Labuzek. 2003. Bacterial degradation and bioremediation of polycyclic aromatic hydrocarbons. Polish J. Env. Studies 12: 12–25.

Nishio, M. 1996. Microbial fertilizers in Japan. ASPAC, Food & Fertilizer Technology Center. Extension Bulletin 430: 13.

Noble, R. 2011. Risks and benefits of soil amendment with composts in relation to plant pathogens. Australasian Plant Pathol. 40: 157–167.

Novak, J.M., W.J. Busscher, D.W. Watts, D.A. Laird, M.A. Ahmedna, and M.A.S. Niandou. 2010. Short-term CO_2 mineralization after additions of biochar and switchgrass to a Typic Kandiudult. Geoderma 154: 281–288.

Ogawa, M. and Y. Yamabe. 1986. Effects of charcoal on VA mycorrhizae and nodule formation of soybeans. Bulletin of the Green Energy Programme Group II, No. 8. Ministry of Agriculture, Forestry and Fisheries, Japan, pp. 108–133.

Ogawa, M. 2007. Reviving pine tree with charcoal and mycorrhiza. Tsukiji Shokan, Tokyo [in Japanese].

Ogawa, M. and Y. Okimori. 2010. Pioneering works in biochar research. Japan. Aust. J. Soil Res. 48: 489–500.

O'Neill, B., J. Grossman, M.T. Tsai, J.E. Gomes, J. Lehmann, J. Peterson, E. Neves, and J.E. Thies. 2009. Bacterial community composition in Brazilian Anthrosols and adjacent soils characterized using culturing and molecular identification. Microb. Ecol. 58: 23–35.

Ottow, J.C. 2011. Mikrobiologie von Böden. Biodiversität, Ökophysiologie und Metagenomik. Springer, Berlin, Germany.

Périé, F.H. and M.H. Gold. 1991. Manganese regulation of manganese peroxidase expression and lignin degradation by the white rot fungus *Dichomitus squalens*. Appl. Env. Microbiol. 57: 2240–2245.

Pietikainen, J., O. Kiikkila, and H. Fritze. 2000. Charcoal as a habitat for microbes and its effect on the microbial community of the underlying humus. Oikos 89: 231–242.

Pignatello, J.J., S. Kwon, and Y. Lu. 2006. Effect of natural organic substances on the surface and adsorptive properties of environmental black carbon (char): Attenuation of surface activity by humic and fulvic acids. Environ. Sci. Technol. 40: 7757–7763.

Püttsepp, Ü., A. Rosling, and A.F.S. Taylor. 2004. Ectomycorrhizal fungal communities associated with *Salix viminalis* L. and *S. dasyclados* Wimm. clones in a short-rotation forestry plantation. Forest Ecol. Managem. 196: 413–424.

Retan, G.A. 1915. Charcoal as a means of solving some nursery problems. Forestry Quarterly 13: 25–30.

Rigamonte, T.A., V.S. Pylro, and G.F. Duarte. 2010. The role of mycorrhization helper bacteria in the establishment and action of ectomycorrhizae associations. Brazilian J. Microbiol. 41: 832–840.

Rillig, M.C., M. Wagner, M. Salem, P.M. Antunes, C. George, and H.-G. Ramke. 2010. Material derived from hydrothermal carbonization: Effects on plant growth and arbuscular mycorrhiza. Appl. Soil Ecol. 45: 238–242.

Ritz, K. and I.M. Young. 2009. The architecture and biology of soils. Life in inner space. School of Environmental & Rural Science, Armidale, Australia.

Ritz, K. Spatial organisation of soil fungi. pp. 179–202. *In:* Rima B. Franklin and Aaron L. Mills [eds.]. 2007. The Spatial Distribution of Microbes in the Environment. Springer, Netherlands.

Rooney, D.C., K. Killham, G.D. Bending, E. Baggs, M. Weih, and A. Hodge. 2009. Mycorrhizas and biomass crops: opportunities for future sustainable development. Trends Plant Sci. 14: 542–549.

Rossman, A.Y. and M.E. Palm. 2006. Why are Phytophthora and other Oomycota not true fungi? Outlooks Pest Management 17: 217–219.

Saito, M. 1990. Charcoal as a micro-habitat for VA mycorrhizal fungi, and its practical implication. Ecological and applied aspects of ecto- and endomycorrhizal Associations. Agriculture, Ecosystems & Environment 29: 341–344.

Saito, M. and T. Marumoto. 2002. Inoculation with arbuscular mycorrhizal fungi: the status quo in Japan and the future prospects. Plant Soil 244: 273–279.

Schimmelpfennig, S. and B. Glaser. 2012. One step forward toward characterization: Some important material properties to distinguish biochars. J. Env. Qual. in press, doi:10.2134/jeq2011.0146.

Scholthof, K-B.G. 2006. The disease triangle: pathogens, the environment and society. Nat. Rev. Micro. 5: 152–156.

Schüßler, A. 2001. A new fungal phylum, the Glomeromycota: phylogeny and evolution. Mycol. Res. 105: 1413–1421.

Schüßler, A. Struktur, Funktion und Ökologie der arbuskulären Mykorrhiza. *In*: Andreas Bresinsky. [eds.]. 2009. Ökologische Rolle von Pilzen. Rundgespräch am 23. März 2009 in München, 37: 97–108. Pfeil Verlag, München, Germany.

Schüßler, A. and C. Walker. 2011. Evolution of the 'plant-symbiotic' fungal phylum, Glomeromycota. The Mycota 14: 163–185.

Schwantes H.O. 1996. Biologie der Pilze. Verlag Eugen Ulmer, Stuttgart, Germany.

Singh, L.P., S. Singh Gill, and N. Tuteja. 2011. Unraveling the role of fungal symbionts in plant abiotic stress tolerance. Plant Sign. & Behav. 6: 175–191.

Six, J., S.D. Frey, R.K. Thiet, and K.M. Batten. 2006. Bacterial and fungal contributions to carbon sequestration in agroecosystems. Soil Sci. Soc. Am. J. 70: 555–569.

Smith, S.E. and D.J. Read. 2008. Mycorrhizal Symbiosis. Academic Press, Amsterdam, Boston.

Solaiman, Z.M., P. Blackwell, L.K. Abbott, and P. Storer. 2010. Direct and residual effect of biochar application on mycorrhizal colonization, growth and nutrition of wheat. Austr. J. Soil Res. 48: 546–554.

Steinbeiss, S., G. Gleixner, and M. Antonietti. 2009. Effect of biochar amendment on soil carbon balance and soil microbial activity. Soil Biol. Biochem. 41: 1301–1310.

Steiner, C., W.G. Teixeira, J. Lehmann, and W. Zech. 2004. Microbial response to charcoal amendments of highly weathered soils and Amazonian Dark Earths in central Amazonia. *In*: B. Glaser and W.I. Woods: Amazonian Dark Earths: Explorations in Space and Time. Springer, Berlin, Germany.

Talbot, J.M., S.D. Allison, and K.K. Treseder. 2008. Decomposers in disguise: Mycorrhizal fungi as regulators of soil C dynamics in ecosystems under global change. Functional Ecol. 22: 955–963.

Tu, C., F.L. Booker, D.M. Watson, X. Chen, T.W. Rufty, W. Shi, and S.J. Hu. 2006. Mycorrhizal mediation of plant N acquisition and residue decomposition: impact of mineral N inputs. Global Change Biology. 12: 793–803.

Wardle, D.A., M.-C. Nilsson, and O. Zackrisson. 2008. Fire-derived charcoal causes loss of forest humus. Science 320: 629.

Warnock, D.D., J. Lehmann, T.W. Kuyper, and M.C. Rillig. 2007. Mycorrhizal responses to biochar in soil—concepts and mechanisms. Plant Soil 300: 9–20.

Warnock, D.D., D.L. Mummey, B. McBride, J. Major, J. Lehmann, and M.C. Rillig. 2010. Influences of non-herbaceous biochar on arbuscular mycorrhizal fungal abundances in roots and soils: Results from growth-chamber and field experiments. Applied Soil Ecology 46: 450–456.

Wengel, M., E. Kothe, C.M. Schmidt, K. Heide, and G. Gleixner. 2006. Degradation of organic matter from black shales and charcoal by the wood-rotting fungus Schizophyllum commune and release of DOC and heavy metals in the aqueous phase. Sci. Total Environ. 367: 383–393.

Yang, Y., G. Sheng, and M. Huang. 2006. Bioavailability of diuron in soil containing wheat-straw-derived char. Sci. Total Environ. 354: 170–178.

Yin, B., D. Crowley, G. Sparovek, W.J. De Melo, and J. Borneman. 2000. Bacterial functional redundancy along a soil reclamation gradient. Appl. Env. Microbiol. 66: 4361–4365.

Zavalloni, C., G. Alberti, S. Biasiol, G.D. Vedove, F. Fornasier, J. Liu, and A. Peressotti. 2011. Microbial mineralization of biochar and wheat straw mixture in soil: A short-term study. Appl. Soil Ecol. 50: 45–51.

Zheng, W., M. Guo, T. Chow, D.N. Bennett, and N. Rajagopalan. 2010. Sorption properties of greenwaste biochar for two triazine pesticides. J. Haz. Mat. 181: 121–126.

Zimmerman, A.R., B. Gao, and M.-Y. Ahn. 2011. Positive and negative carbon mineralization priming effects among a variety of biochar-amended soils. Soil Biol. Biochem. 43: 1169–1179.

4

The Potential of Biochar Amendments to Remediate Contaminated Soils

Jose L. Gomez-Eyles,[1,*] Luke Beesley,[2] Eduardo Moreno-Jimenez,[3] Upal Ghosh[4] and Tom Sizmur[5]

Introduction

[1]Department of Chemical, Biochemical, and Environmental Engineering, University of Maryland Baltimore County, Baltimore, Maryland 21250, United States; E-mail: jlge@umbc.edu

[2]The James Hutton Institute, Craigiebuckler, Aberdeen, AB15 8QH, UK; E-mail: Luke.Beesley@hutton.ac.uk

[3]Department of Agricultural Chemistry, Universidad Autónoma de Madrid, 28049 Madrid, Spain; E-mail: eduardo.moreno@uam.es

[4]Department of Chemical, Biochemical, and Environmental Engineering, University of Maryland Baltimore County, Baltimore, Maryland 21250, United States; E-mail: ughosh@umbc.edu

[5]Department of Materials Science and Engineering, Iowa State University, 2220 Hoover Hall, Ames, Iowa 50010, United States; E-mail: tosizmur@mail.iastate.edu

*Corresponding author

Introduction

There is a legacy of polluted soils worldwide, contaminated with a variety of different chemicals from a wide range of industrial (e.g., electricity generation, oil refining, mining), agricultural (e.g., pesticide application) and urban (e.g., waste disposal, motor vehicle discharges) sources. Soils are considered polluted when they have an excess of an element or compound which, through direct or indirect exposure, causes a toxic response to biota resulting in unacceptable risks to the environment or human health (Adriano 2001, Abrahams 2002, Vangronsveld et al. 2009). This is a cause for concern in many countries such as the U.S where over 100,000 contaminated sites have been identified (Connell 2005), or the E.U. member states that have reported 250,000 polluted sites that need urgent remediation (Mench et al. 2010). The remediation of contaminated soils is therefore receiving increasing attention from governments, legislators, industries and the general population. Due to the wide range and dispersive nature of many contaminant sources, soil contamination is often too widespread for *ex situ* remediation options (e.g., excavating and burying polluted soils in landfill sites) to be practically, environmentally or financially viable. This has resulted in the development of more sustainable, *in situ* remediation treatments that, in many cases, involve the application of amendments directly to contaminated soils. For soils contaminated with inorganic pollutants like heavy metals, these amendments include clay minerals, zeolites, lime or composts (Simon 2000, Mench et al. 2003, Vangronsveld et al. 2009), whereas for organic contaminants like polycyclic aromatic hydrocarbons (PAHs), or polychlorinated biphenyls (PCBs), activated carbons have been favored (Brändli et al. 2008, Ghosh et al. 2011). The application of this kind of *in situ* amendments does not remove the contaminants from the soil, so their success is based on a reduction in the bioavailability and/or mobility of the contaminants in question. This reduction is achieved by altering the physico-chemical and biological characteristics of the soil, ultimately reducing the risk of contaminant uptake by fauna and flora, or leaching into waters. Biochars have received interest recently as an amendment for soil remediation purposes due to their potential to reduce the bioavailability of organic contaminants (Gomez-Eyles et al. 2011, Hale et al. 2011), inorganic contaminants (Beesley and Marmiroli 2011), and both organic and inorganic contaminants simultaneously (Cao et al. 2009, Beesley et al. 2010b). This is demonstrated by a rapid increase in the number of studies published featuring, within their title, the words 'biochar', 'soil', and 'contaminated' or 'polluted' from just 1 in 2007 to almost 30 in 2011 (Fig. 1).

This increase reflects a growing awareness and use of biochars for the remediation of contaminated soils. In this chapter the efficacy of biochars for contaminant mitigation in soils is examined, describing the mechanisms by which biochars can reduce contaminant mobility and bioavailability. We also identify possible complications that could arise from amending contaminated sites with biochars and discuss new developments in biochar technology and production to overcome these complications.

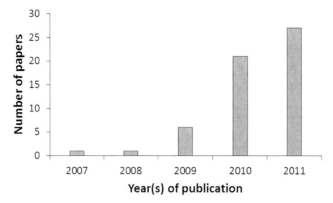

Figure 1. Published studies (ISI) featuring a title with the words 'biochar' and 'soil' and either 'contaminated' or 'polluted'.

1 The Remediation of Soils Contaminated with Organic Pollutants using Biochars

1.1 Introduction

Research into the use of biochars to remediate soils polluted with organic contaminants is still in its early stages. Only a few studies to date have investigated the use of biochar within this context (Beesley et al. 2010b, Bushnaf et al. 2011, Gomez-Eyles et al. 2011, Hale et al. 2011, Chai et al. 2012). However, a large number of studies have documented a reduction in the bioavailability of organic pesticides following biochar or charcoal application in agricultural soils (Yang and Sheng 2003, Yang et al. 2006, Yu et al. 2006, Cao et al. 2009, Spokas et al. 2009, Yu et al. 2009, Wang et al. 2010, Yu et al. 2011). This reduction in bioavailability occurs due to the high affinity of hydrophobic organic compounds for black carbon-like materials such as biochar. Black carbons are produced by the partial combustion or pyrolysis of biomass and fossil fuels, and are present in soils worldwide due to both natural (e.g., forest fires or volcanic eruptions), and anthropogenic processes (e.g., fossil fuel burning). The enhanced sorption of hydrophobic organic compounds to black carbon-like materials (Jonker and Koelmans 2002, Ghosh et al. 2003, Cornelissen and Gustafsson 2005, Cornelissen et al. 2005, Ghosh 2007), can significantly reduce the concentration of these contaminants in the aqueous phase (Cornelissen et al. 2006a). Reductions in pore water concentrations can be hugely beneficial in terms of reduced leaching and dispersion rates, but also because the aqueous concentration of organic contaminants is closely linked to the contaminant fraction available for uptake by soil biota (Jonker et al. 2007, Gomez-Eyles et al. 2012).

The reduction in the bioavailability of organic contaminants due to the presence of black carbon (Ghosh et al. 2003), has prompted the idea of deliberately introducing clean types of these black carbon-like materials into soil and sediments. Initial research has demonstrated that the application of activated carbon (a black carbon produced by the high temperature pyrolysis of different precursor materials followed by an activation step to maximize its sorption capacity) can substantially reduce the aqueous concentrations, and hence the bioaccumulation, of organic compounds in sediments (Zimmerman et al. 2004, Cho et al. 2007, Sun and Ghosh 2008, Cho et al. 2009, Beckingham and Ghosh 2011, Ghosh et al. 2011). Activated carbons have a sorption capacity for organic contaminants (Hale and Werner 2010) that exceeds that of other black carbon materials usually present in soils like soot, chars or coal (Jonker and Koelmans 2002), and some studies have shown that activated carbon amendments can also reduce organic contaminant bioavailability in soils (Brändli et al. 2008, Fagervold et al. 2010, Paul and Ghosh 2011). Biochars could therefore be used in a similar manner for the *in situ* remediation of soils contaminated with organic compounds.

Despite their sorption capacity for hydrophobic organic compounds not being as high as that of activated carbons, biochars have a number of advantages over activated carbons. Biochars are generally produced at lower pyrolysis temperatures (200–700°C) than activated carbons (800–2000°C), and their production does not usually include an activation step (such as purging with steam or acid). This makes their manufacture cheaper, less energy intensive, and hence, more attractive from a carbon sequestration viewpoint.

Due to the potential for biochar to be applied to agricultural land, to improve soil quality and sequester carbon, most studies to date investigating the effect of biochars on organic compound bioavailability have focused on organic pesticides. Studies have shown that the sorption of pesticides like diuron or terbuthylazine in soils amended with different levels of biochars increased by factors of between 2.7 and 125 (Yang and Sheng 2003, Yang et al. 2006, Yu et al. 2006, Wang et al. 2010). This increased sorption has been shown to reduce the rate of microbial degradation of diuron in soil by >10% and of chlorpyrifos and carbofuran by >40% (Yang et al. 2006, Yu et al. 2009). This confirms the ability of biochar amendments to reduce organic compound bioavailability; importantly, this reduction in bioavailability has been shown to be accompanied by a reduction in total plant residues of chlorpyrifos (10%) and carbofuran (25%) (Yu et al. 2009). Similarly, the sorption coefficient of soils contaminated with other organic pollutants like polycyclic aromatic hydrocarbons (PAHs), have been found to increase by factors of up to 700 after the addition of biochars (Zhang et al. 2010). This increased sorption translates into reductions in the bioavailable PAH fraction of >40% (Beesley et al. 2010b) and up to 45% reductions in PAH accumulation in earthworms exposed to biochar amended soils (Gomez-Eyles et al. 2011). However, it is important to realize these reductions in bioavailability are still lower than those reported after activated carbon amendments to sediments (Zimmerman et al. 2004, Millward et al. 2005, Cho et al. 2009).

1.2 Mechanisms for Biochars' Sorption of Organic Contaminants

The sorption of organic contaminants to solid surfaces in the soil matrix can greatly reduce their bioavailability. Organic compounds can sorb to biochars by (i) linear and non-competitive absorption or partitioning (Accardi-Dey and Gschwend 2001, Chun et al. 2004, Zhou et al. 2009), and by (ii) non-linear extensive and competitive surface adsorption (Cornelissen et al. 2005) (Fig. 2). This dual-mode sorption concept was originally developed to describe organic contaminant sorption to organic matter (Pignatello and Xing 1995, Xing and Pignatello 1997), but it can be applied to biochars as they

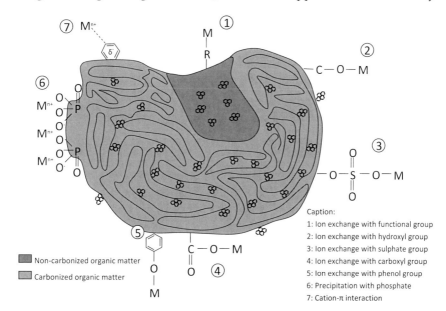

Caption:
1: Ion exchange with functional group
2: Ion exchange with hydroxyl group
3: Ion exchange with sulphate group
4: Ion exchange with carboxyl group
5: Ion exchange with phenol group
6: Precipitation with phosphate
7: Cation-π interaction

Figure 2. Mechanisms of organic (e.g., PAHs) and inorganic (M =metal) contaminant sorption to biochar. Organics either partition into the non-carbonised organic matter or adsorb onto the internal surfaces of the carbonized organic matter. Inorganics bind to different functional groups on the biochar surface and can also adsorb to the biochar via cation-π interactions.

are generally not fully carbonized and are therefore made up of different proportions of carbonized and non-carbonized organic matter (Chen et al. 2008). Biochars that are produced at low pyrolysis temperatures contain a higher degree of non-carbonized organic matter, so organic contaminant sorption to them is a combination of both partitioning and adsorption. With increasing pyrolysis temperatures, biochars become more carbonized and the sorption processes will be dominated by the adsorption of the organic contaminants to the internal biochar surface (Chun et al. 2004). However, at high contaminant concentrations, non-linear competitive adsorption sites may become saturated, making the partitioning phase gain importance (Chen and Huang 2011), as discussed in section 1.4.

The exact adsorption mechanism by which non-polar (hydrophobic) organic contaminants like PAHs sorb to carbonized organic matter has not yet been unraveled (Cornelissen et al. 2005). However, it is clear that this kind of sorption generally occurs on the internal surface regions of the carbon as demonstrated by (i) microscopic observations (Ghosh et al. 2000), and (ii) good correlations between surface areas and sorption coefficients (Bucheli and Gustafsson 2000). One of the adsorption mechanisms hypothesized is π-π interactions, between the π-electrons within the aromatic rings of hydrophobic organic compounds, like PAHs, and, the π-electrons in the aromatic planes of the carbonized organic matter (Zhu et al. 2005, Chen and Huang 2011). This may be one of the reasons why the sorption coefficients of planar hydrophobic organic compounds to black carbon-like materials can be up to an order of magnitude higher than non-planar ones with otherwise similar partitioning properties, as discussed by Cornelissen et al. (2005).

It is important to note that most of the discussion on organic contaminants in this chapter focusses on non-polar hydrophobic contaminants (e.g., PAHs, organophosphate and organochlorine pesticides, dioxins, furans etc.), but other sorption mechanisms are important for more polar organic compounds (e.g., nitrobenzenes, pyridazinone herbicides, pharmaceuticals etc.). The presence of polar functional groups on the surface of biochars play a bigger role in the sorption of these more polar contaminants (Sun et al. 2011), as with inorganic contaminants. These sorption mechanisms are discussed in more detail in section 2.2.

1.3 Optimizing Biochar Production for Organic Contaminant Sorption

Although the exact mechanism for organic compound adsorption to biochars has not been confirmed, research to date clearly indicates that to optimize the sorption of non-polar hydrophobic organic contaminants it is necessary to maximize the surface area of the biochar. The most immediately obvious way to do this is to increase the pyrolysis temperature at which the biochar is produced because this helps to develop biochar microstructure. A number of studies have shown that increasing the pyrolysis temperature of biochars increases their capability to adsorb hydrophobic organic pesticides (Yu et al. 2006, Wang et al. 2010, Zhang et al. 2010), reducing their uptake by plants (Yu et al. 2009, Yang et al. 2010). A more detailed study has shown that increasing pyrolysis temperature increases the surface area of pine needle-derived biochars from 0.65 $m^2 g^{-1}$ at 100°C to 490.8 $m^2 g^{-1}$ at 700°C (Chen et al. 2008) (Fig. 3).

It is also possible to increase biochar surface area by introducing an extra activation step during manufacture, as in the production of activated carbons. The physical activation of biochars can be carried out by gasification using carbon dioxide or steam at relatively high temperatures (600–900°C) to develop their meso and microporosity (Azargohar and Dalai 2008). Biochars can also be activated chemically by soaking the precursor material in activating agents like phosphoric acid or zinc chloride before pyrolysis (Jaramillo et al. 2009, Al-Qaessi and Abu-Farah 2010, Lim et al. 2010). The chemical agent is thought to dehydrate the sample inhibiting tar formation and volatile compound evolution during the pyrolysis process, increasing its surface area (Williams and Reed 2004). Activated biochars generally

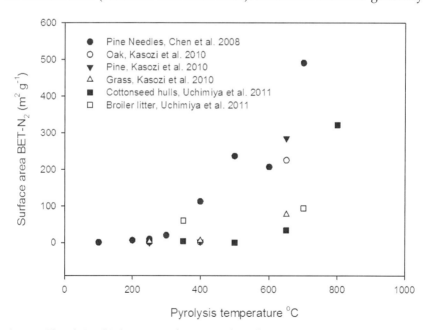

Figure 3. The relationship between surface area and pyrolysis temperature of biochars derived from different source materials.

have surface areas of between 1000 and 2000 $m^2 g^{-1}$ (Lim et al. 2010). This is substantially higher than the surface areas of between 250 and 500 m^2 g^{-1} usually reported for non-activated biochars produced at high pyrolysis temperatures (Chen et al. 2008, Kasozi et al. 2010). Activated biochars pyrolyzed at high temperatures therefore have the highest non-polar organic contaminant remediation potential. However, activated biochars are less economically and environmentally sustainable than unactivated biochars because (i) their manufacturing process is more expensive, (ii) there is more carbon loss during the activation process, and (iii) a higher input of energy is required. These factors reduce their potential to sequester carbon.

It is also important to consider that the choice of precursor materials will also have an effect on the sorption capacity of different biochars, especially for polar and ionizable molecules. A study has shown that the sorption capacity for catechol, a small phenolic compound, by biochars produced under the same pyrolysis conditions but from different source materials, varied by up to an order of magnitude (Kasozi et al. 2010). However, changes in sorption capacity were much more dramatic (more than 2 orders of magnitude) when pyrolysis temperature was changed than when biomass type was changed.

It is important to note that sorption mechanisms are different when considering polar organic compounds, so optimizing biochar production by increasing the surface area may not be effective for the remediation of soils contaminated with polar organic contaminants. In fact, when considering polar organic contaminants, surface area is a minor factor controlling sorption. Instead, biochar polarity controls sorption (Sun et al. 2011). Reducing pyrolysis temperatures to increase biochar functionality instead of increasing them to increase the biochar surface area would therefore be more effective when attempting to optimize biochar sorption properties for the more polar organic contaminants. Again, this is similar to optimization for inorganic contaminants, as discussed in section 2.2.

The particle size of a biochar can also have a large impact on its effectiveness. Although Hale et al. (2011) showed that grinding biochar to a smaller particle size did not affect hydrophobic organic contaminant sorption (implying that sorption processes occur at the subparticle scale), laboratory and field studies have demonstrated reduced effectiveness of activated carbon amendments with increasing particle size (Zimmerman et al. 2005, Chai et al. 2012). Contaminant mass transfer is slow within biochar particles, so finer granular biochars that can be more evenly spread in the soil are more effective for remediation purposes, although their deployment into soils may be more challenging.

1.4 Possible Complications

1.4.1 Long Term Effectiveness

As research into the use of biochars for organic contaminant remediation purposes is still in its early stages, there is no long term field data enabling an assessment of the historical effectiveness of biochars to sequester organic contaminants in soils. However, some laboratory data do indicate that biochar performance may decrease with time. As mentioned in section 1.2, adsorption of non-polar organic compounds to biochars is non-linear, suggesting that even if biochars with high surface areas are produced, sorption sites could become saturated. The blocking of sorption sites in activated carbons by organic matter has been extensively reported (Kilduff and Wigton 1998, Cornelissen and Gustafsson 2004, Rhodes et al. 2010), and several studies have hypothesized that this is the likely cause for the diminished capability of aged biochar to adsorb organic contaminants (Yang and Sheng 2003a, Zhang et al. 2010). Conversely, a recent study attempting to assess the long term effectiveness of biochars and activated carbons, by artificially aging amended soils in the laboratory by different methods, revealed more encouraging results (Hale et al. 2011). Although the authors note that two months of harsh laboratory aging cannot replicate the natural environment, they report a limited effect on the sorption capacity of biochar and activated carbon by aging or by the presence of soil. They also found the effects of aging were greater on the biochar than the activated carbon, presumably due to less sorption sites being available on the surface of the biochar.

It is also possible for the performance of carbonaceous sorbents to drop due to competition between contaminants (Yang and Sheng 2003b, Cao et al. 2009), suggesting that biochars may be more effective in soils with low levels of contamination. Competition effects would be less prevalent when using low temperature pyrolysis biochars as the partitioning domain prevails over the non-linear competitive adsorption one (see section 1.2). In a study comparing activated carbon to biochars produced at lower temperatures, the biochars were found to absorb Pb and atrazine linearly into their non-carbonized organic matter (Cao et al. 2009). This may have its advantages when remediating soils with contaminants that compete for adsorption sites. However, the sorption capacities of the low temperature biochars were an order of magnitude lower than that of the activated carbon for atrazine.

Long term field trials still need to be carried out to fully assess the potential of carbonaceous sorbents to remediate soils contaminated with organic pollutants. A site specific compromise between the increased sorption capacities of high pyrolysis temperature activated chars and the benefits provided by low temperature chars in terms of carbon sequestration, reduced toxicity and linear sorption is needed, depending on remediation and restoration goals.

1.4.2 Reduced Herbicide Efficiency

From an agricultural perspective, increased sorption and reduced bioavailability of organic pesticides to soil can be beneficial in terms of reducing pesticide residues in crops (Yu et al. 2009). However, this decreased bioavailability can also be detrimental in terms of reducing herbicide efficiency, resulting in the need for higher application rates of these chemicals (Yu et al. 2006, Spokas et al. 2009, Kookana 2010, Nag et al. 2011). It is therefore important for farmers to understand to what extent biochars will reduce the efficacy of the specific herbicide they are using in their soil and adjust its application accordingly, if regulation permits. The potential economic and environmental disadvantages of having to apply more herbicide should then be weighed against the potential benefits of biochar application.

1.4.3 Reduced Contaminant Degradation

Although amending soils with biochars will reduce the bioavailability and hence the toxicity of organic compounds, it is also likely for them to reduce the rate at which organic compounds are degraded in the environment precisely due to this reduction in bioavailability. A number of studies have shown that the presence of biochar or black carbons can reduce the degradation of organic pesticides (Yang et al. 2006, Yu et al. 2009) and other organic contaminants (Talley et al. 2002, Ghosh et al. 2003, Rhodes et al. 2008, Yang et al. 2009). This should be considered when amending soils with biochar, especially when soils contain organic compounds that would otherwise degrade relatively rapidly. Soils that are contaminated with petroleum hydrocarbons due to oil spills or leakages from storage tanks, for example, will have a range of organic contaminants with different properties. It is likely that more recalcitrant PAHs will also be present in these soils and reducing their bioavailability using carbon amendments has been shown to be effective, as described in sections 1.1 and 1.2. However, this could compromise the degradation of the more readily degradable hydrocarbons. A recent study has investigated this issue and found that despite increased sorption of the monoaromatic hydrocarbons, microbial respiration was comparable in biochar amended and non-amended petroleum hydrocarbon contaminated soils (Bushnaf et al. 2011). The authors concluded that total petroleum hydrocarbon degradation was limited by a factor other than hydrocarbon substrate availability and actually found that the reduced availability of monoaromatic hydrocarbons in the biochar amended soil led to a greater biodegradation of the other petroleum compounds like cyclic or branched alkanes. Despite these findings, enhanced natural attenuation strategies may be more effective than biochar application for soils polluted with more readily degradable contaminants. Biologically active carbon amendments could be more suited to soils with these less persistent contaminants. Impregnating biochars with a biofilm of hydrocarbon degrading bacteria could be a way of addressing this issue, as has been achieved with activated carbons (Leglize et al. 2008). The biochar particles would supply a surface for the biofilm to form around, which in turn can provide a protective environment for the microbes to live in. The high affinity of the hydrophobic organic contaminants to the biochars could then enable the concentration of the hydrocarbon substrate at sites in close proximity to the degrading bacteria. However, it is premature to say whether biologically active biochar amendments would be an effective *in situ* soil remediation strategy.

1.4.4 Possible Toxic Effects to Soil Fauna

The toxicity of biochar to soil organisms is already discussed in more detail in other chapters within this book, but briefly, activated carbon amendments, especially in a powdered form, have in some occasions been found to have a negatively affect some sediment and soil organisms, especially those that live in intimate contact, and ingest large volumes, of sediment and soil (Jonker et al. 2009, Fagervold et al. 2010). It is believed that this occurs due to the strongly sorbing nature of the powdered activated carbon reducing the amount of soluble organic carbon available for uptake by soil and sediment dwelling organisms (Jonker et al. 2009). However, other studies of activated carbon amendments in the granular form, directly incorporated into sediments, observed little impacts on benthic organisms (Cornelissen et al. 2006b, Sun and Ghosh 2008). Importantly, field studies have demonstrated no impact on the benthic community in activated carbon treated sediments (Cho et al. 2009). Weight loss effects have also been found in earthworms after amending soils with biochars (Gomez-Eyles et al. 2011), but this effect was not apparent in other studies of earthworms in biochar-amended soils (Wen et al. 2009, Liesch et al. 2010). The reduced sorption capacity of biochars, relative to activated carbon, might be therefore be advantageous when considering the toxicity of biochar to soil-dwelling organisms, although this has yet to be fully demonstrated.

2 The Remediation of Soil Contaminated with Inorganic Pollutants using Biochars

2.1 Introduction

Research surrounding the use of biochar to remediate soils with inorganic pollutants is slightly more advanced than for organic pollutants, with studies ranging from aqueous sorption assays to mesocosm trials. However, there is still a lack of long-term field data. Studies where biochars are added to aqueous solutions containing inorganic pollutants are good model systems to determine the direct impacts of biochar on inorganic contaminant mobility because interactions between the soil matrix and the biochar (indirect effects), or soil dilution effects cannot complicate the interpretation of results. For Cu and Zn, sorption to hardwood and corn straw-derived biochars has been determined in batch testing (Chen et al. 2011b). Between 77–83% of the total metal sorption occurred within two hours of biochar application, but although the total metal sorption increased with increasing amounts of char added to solution, the efficiency (i.e., mg metal removed per g biochar) decreased with increasing biochar loading. A suggested reason was that aggregation of biochar particles may have occurred in the water with greater biochar loading, reducing the surface area and decreasing the effective sorption sites. However, Mohan et al. (2012) demonstrated that chars can swell in water, increasing their surface area, so it is not altogether clear whether adding biochars to aqueous systems would result in a decrease or increase in their efficiency over time. Liu and Zhang (2009) assessed the Pb sorption potential of pinewood- and rice husk-derived biochars produced by hydrothermal liquefaction at 300ºC and how this was affected by pH and equilibration time. The optimal equilibration time was found to be ~5h, and optimum Pb removal at pH 5. At low concentrations, Pb ions were located on the outer surfaces of the biochar, but at greater solution concentrations Pb entered the inner structure illustrating the importance of the physical structure of the biochar in element retention.

Leaching column tests, where a contaminated solution flows through a solid (biochar, soil or a combination) matrix, brings batch leaching tests a step closer to field conditions by introducing a temporal assessment of the effects of the leached media. Beesley and Marmiroli (2011) used a 9 week flow-through leaching column experiment to test the immobilization of Cd and Zn by a hardwood-derived biochar, finding 300 and 45 fold reductions in leachate Cd and Zn concentrations respectively. Also of note was the reduction in water-extractable carbon during the time course of this test; complexation of pollutants with soluble fractions of carbon can affect their transport and retention in soils, which is a complicating factor of adding carbon-rich amendments to soils.

The use of pot, microcosm or mesocosm trials to assess biochar application to contaminated soils, either in the laboratory or in the field, brings us another step closer to 'real world' field conditions, but allowing controlled variation. Experiments using 'real' contaminated soils allow the testing of additional effects that may have been overlooked in column or batch laboratory studies. In such an experiment, amending contaminated mine tailings with 0–10% (by weight) of an orchard prune-derived biochar decreased extractable Cd, Pb and Zn concentrations (Fellet et al. 2011). Beesley et al. (2010) extracted soil pore water over 60 days from a metal contaminated soil amended with a hardwood-derived biochar. Biochar reduced Cd and Zn in pore water and increased ryegrass shoot elongation when extracted pore water was used in a petri-dish phyto-toxicity test. Another study found decreases in Zn concentrations in soil leachate after pecan shell-derived biochar addition to an acidic soil (Novak et al. 2009a). Leachate concentrations of Zn, Cu, P, K, Mg and Ca have also been shown to decrease with increasing biochar application rates (Laird et al. 2010). Cao et al. (2009) found that a dairy manure derived biochar retained more Pb than activated carbon, despite having a much lower surface area. The authors suggested that biochar reduced Pb mobility due to the precipitation of insoluble Pb-phosphates (Fig. 2). Biochars derived from manure have been reported to be rich in P (Cao and Harris 2010). Other authors have reported that biochar addition reduced pore water Pb concentrations to half of their original value (Karami et al. 2011). Ahmad et al. (2012) determined the effects of a biochar amendment on Pb in a shooting range soil, finding that biochar addition gave a ~75% reduction in the bioavailable (water extractable) Pb concentration and a ~12% reduction in bioaccessible (exchangeable) Pb concentration. It therefore seems that biochars are able to reduce the mobility of a range of inorganic elements quite effectively.

As discussed in section 2.2, biochars have some indirect effects on soil chemistry that may impact on the mobility of elements in soil. One such effect is the impact of biochar addition on soil pH. The addition of biochars which typically have a pH of 7–9, to acidic soils, results in an increase in the soil pH. Nine biochars (pH 6.4–10.4) were incubated for 60 days in an acidic soil (pH 4.3) finding that the resulting increase in soil pH was positively correlated with the starting pH of the biochar, so it was possible to predict the liming effect of biochars before they were added (Yuan and Xu 2011). Karami et al. (2011) and Beesley and Marmiroli (2011) also noted increased pH after the addition of biochar. Generally an increase in pH would be expected to increase arsenic (As) mobility because raising soil pH increases the mobility of the arsenic oxyanion. Namgay et al. (2010) noted increases of phosphate-extractable As, whilst Hartley et al. (2009) found an increase in pore water As after biochar addition. Co-mobilisation of inorganic contaminants with dissolved organic carbon (DOC) can increase their mobility in soils after the addition of organic amendments (Bernal et al. 2007). Concentrations of 0.1–109 g kg^{-1} water soluble organic carbon in biochars with a range of source materials and production temperatures have been reported (Gell et al. 2011). Beesley et al. (2010) found that Cu mobility increased in a contaminated soil after hardwood-derived biochar amendment because biochar increased DOC in soil pore water. Gomez-Eyles et al. (2011), however, showed a decrease in water soluble organic carbon and a concurrent decrease in Cu mobility after amendment of a hardwood-derived biochar. The effect of biochar on co-mobilisation of Cu with DOC may therefore be soil and biochar specific.

Within functional soil ecosystems, soil animals play an important role in the interaction between soil amendments and the native soil fauna. Furthermore, soil fauna may modify the effects of biochar on microbial activity, or assist biochar's incorporation into bulk soil (Lehmann et al. 2011). *Lumbricus terrestris* earthworms reduced the concentration of DOC in the pore water of a biochar-amended soil, which reduced pore water concentrations of As, Cu and Pb (Beesley and Dickinson 2010). A number of mechanisms were proposed for these effects; biochar's liming effect on soil, co-mobilisation of metals and As with DOC, and the ability (or inability) of earthworms to ingest biochar. In the case of the latter, this could influence the surface area of the biochar, and, since biochar has been found to adsorb DOC (Pietikäinen et al. 2000), then an increased surface area through fragmentation by earthworms, could reduce mobile DOC. However, in a more recent study biochar was found to considerably decrease the concentrations of water soluble Cu, Pb and Zn in a contaminated mine waste soil but *L. terrestris* had no effect on concentrations of water soluble metals (Sizmur et al. 2011), so the effect of earthworms on the fate of metals in biochar amended soils is not clear.

2.2 Mechanisms for Biochars' Immobilization of Metals

Organic amendments immobilise metals by different physico-chemical mechanisms which reduce contaminant mobility and bioavailability (Bolan and Duraisamy 2003, Bernal et al. 2007). The pyrolysis of organic matter to produce biochar results in a material with a number of remarkable properties such as (i) a high internal surface area, (ii) a high cation exchange capacity (CEC), (iii) a high pH, and (iv) a low rate of degradation. As described in section 1.3, the surface area of a biochar can be increased by increasing the pyrolysis temperature or by activation. Increasing surface area can be beneficial for the sorption of hydrophobic organic contaminants, but increasing the pyrolysis temperature and surface area of a biochar often results in a decrease in its capacity to immobilize metals. This decrease has been observed for biochars produced from cottonseed hulls (Uchimiya et al. 2011b), dairy manure (Cao et al. 2009) rice husks (Pellera et al. 2012), olive pomace (Pellera et al. 2012), coconut coir (Shen et al. 2012) and broiler litter (Uchimiya et al. 2010b). Similarly to polar organic contaminants, the immobilization of metals by sorption on biochar surfaces is controlled to a larger extent by the surface chemistry of a biochar rather than by its surface area. The abundance of functional groups, the presence of mineral salts, and the surface charge density play a larger role in determining a biochars' ability to immobilize metals (Fig. 2) than the surface area.

There are five broad mechanisms for metal immobilization by biochar:

1. Increase in soil pH
2. Ion exchange
3. Physical adsorption
4. Precipitation
5. Improvement of soil properties

2.2.1 Increase in Soil pH

As mentioned in section 2.1, the addition of biochar to soils generally increases soil pH (Yamato et al. 2006, Chan et al. 2007, Uchimiya et al. 2010b, Van Zwieten et al. 2010, Bell and Worrall 2011). Biochars with the highest proportions of ash, produced at the highest temperatures, from mineral-rich source materials, result in the greatest increases in soil pH (Cao and Harris 2010, Lehmann et al. 2011). This is unsurprising since the addition of minerals (e.g., lime) to soils is routinely carried out to raise soil pH and increase the availability of plant nutrients such as P, Ca and Mg (although lime can reduce the availability of some micronutrients such as Fe and Zn). By raising the pH of a soil, the mobility and availability of metals decrease due to less competition between H^+ and M^{n+} for sorption sites. In biochar-amended soils, the decrease in metal mobility at a higher pH cannot be explained by the sorption of metals on the surface of biochar alone, there is also an increase in sorption to other soil surfaces. This is because an increase in soil pH also leads to an increase in the pH-dependent CEC of organic matter, clay minerals and oxyhydroxides present in soils. A decrease in metal mobility after raising the pH is also due to the precipitation of metals as insoluble mineral species (hydroxides, phosphates, carbonates, etc.) (Lindsay 1979). Metal immobilization due to increases in pH can only be considered temporary as it is not known how long the elevation of soil pH persists in soils after biochar application.

2.2.2 Ion Exchange

The pyrolysis of organic matter and subsequent exposure to the atmosphere leads to the oxygenation of biochar surfaces (Cheng et al. 2006). This oxygenation results in the formation of oxygen containing functional groups (e.g., carboxyl, hydroxyl, phenol and carbonyl groups) (Fig. 2) over the vast internal surface area of the biochar (Liang et al. 2006, Lee et al. 2010, Uchimiya et al. 2010b, Uchimiya et al. 2011b). These functional groups give rise to a considerable negative charge and a high CEC (Cation Exchange Capacity). The CEC of biochars first increases and then decreases with increasing pyrolysis temperatures (Gaskin et al. 2008, Lee et al. 2010, Harvey et al. 2011, Mukherjee et al. 2011) with a peak CEC of up to 45 cmol$_c$ kg^1 generally occurring between 250 and 350°C (Fig. 4), depending on the source material. The lower CEC observed after higher temperature pyrolysis is concurrent with a lower oxygen:carbon ratio and a decrease in the abundance of oxygenated (acid) functional groups (Cheng et al. 2006, Lee et al. 2010, Harvey et al. 2011, Uchimiya et al. 2011a, Shen et al. 2012), which are likely to be responsible for the high CEC and metal retention in <500°C pyrolysis temperature biochar. Therefore, the considerable capacity for metal immobilization demonstrated by low temperature (<500°C), faster

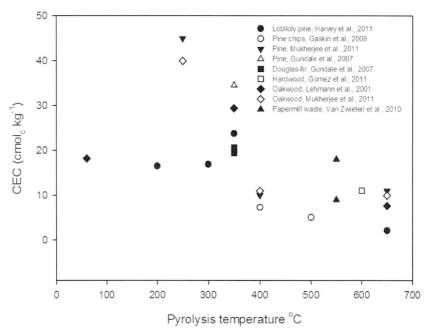

Figure 4. The relationship between cation exchange capacity (CEC) and pyrolysis temperature of hardwood-derived biochars.

pyrolysis biochars (Beesley et al. 2010b, Beesley and Marmiroli 2011) is, at least in part, due to the high CEC of these biochars.

Sorption of metal ions on biochar surfaces occurs concurrently with the release of H (Uchimiya et al. 2010b), but also of Na, Ca, S, K and Mg (Uchimiya et al. 2011a) into solution. This finding not only indicates the sorption of metals on protonated (acid) functional groups but also their exchange with other cations. It is likely that phosphorus- and sulphur-containing ligands play an important role in the sorption of metal ions such as Pb and Hg that have a particular affinity for phosphates and sulphates, respectively (Cao et al. 2009, Uchimiya et al. 2010b). The oxygenated functional groups on the surface of biochars may also play a role in the oxidation of redox sensitive metals. The application of a low pyrolysis temperature (250°C) coconut coir-derived biochar completely reduced Cr(VI) to Cr(III) after adsorption to its surface functional groups (Shen et al. 2012). Higher pyrolysis temperature (350 and 600ºC) coconut coir-derived biochars removed less Cr from solution and reduction occurred before adsorption.

There is evidence to suggest that much of the CEC of biochars pyrolyzed at low temperatures may arise due to the presence of non-carbonized organic matter. This non-carbonized fraction may be a particulate constituent of the biochar (Mukherjee et al. 2011) (Fig. 2), or be sorbed to the surface of biochar, thereby increasing the surface charge density (Liang et al. 2006). Biochars with high proportions of non-carbonized organic matter may increase the concentration of DOC in the soil solution. As discussed in section 2.1, this increase in DOC, released from the biochar, may result in the co-mobilisation of metals with a high affinity for organic carbon (i.e., Cu, Pb, Hg) and lead to greater metal mobility in biochar-amended soils (Beesley et al. 2010b, Uchimiya et al. 2010a).

2.2.3 Physical Adsorption

Several studies reveal that the immobilization of metals by biochar results in a non-stoichiometric release of protons and other cations from the surface of biochars (Uchimiya et al. 2010b, Harvey et al. 2011). Essentially, more metals are adsorbed than protons or cations are released. In addition, sorption can take place at pH levels below the point of zero net charge (Sanchez-Polo and Rivera-Utrilla 2002). This discrepancy indicates that the immobilization of metals by biochar cannot be completely explained by cation or proton exchange alone. Interpretation of the thermodynamic parameters of metal sorption to biochars and activated carbons reveals that sorption is an endothermic physical process (Kannan and Rengasamy 2005, Liu and Zhang 2009, Harvey et al. 2011). That is, an electrostatic interaction occurs between the positively charged metal cations and π-electrons associated with either C=O ligands or (more likely) C=C of a shared electron 'cloud' on aromatic structures in biochar (Swiatkowski et al. 2004, Cao et al. 2009, Uchimiya et al. 2010b, Harvey et al. 2011) (Fig. 2).

When six carbons combine to form a benzene ring, each carbon donates an electron to the structure which is then 'delocalized'. This results in an 'electron cloud' or a π-cloud above and below the planar surface of the benzene ring. Metal cations are positively charged due to 'missing' electron(s) from their d-orbitals. When a positively charged cation approaches the benzene ring, the electron cloud becomes polarized and there is a weak electrostatic interaction between the negatively charged planar surface of the benzene ring and the positively charged metal cation (Fig. 2). The bond energies of cation-π interactions are typically in the range 1 to 30 kcal mol^{-1}(Zarić 2003), while the bond energies of transition metal-carbon bonds are typically an order of magnitude higher (>100 kcal mol^{-1}) (Simoes and Beauchamp 1990).

With increasing pyrolysis temperature the aromaticity of biochars increases while the abundance of oxygenated functional groups decreases (Harvey et al. 2011, McBeath et al. 2011). Therefore, with increasing pyrolysis temperature, there is an increasing proportion of cation sorption due to 'weak' electrostatic bonding (i.e., cation-π interactions) and a decreasing proportion due to relatively stronger chemisorption (i.e., by cation exchange). This indicates that there may be a trade-off between long term and short term stability when selecting a biochar to immobilize metals in soils. Low temperature pyrolysis will likely result in effective short term metal immobilization due to the formation of inner sphere and outer sphere complexes with oxygenated (acid) functional groups. However, these may degrade over time in the soil environment (within the first 90 days after application: Zimmerman et al. 2011), and therefore release the metals back into solution. High temperature pyrolysis will result in a negative surface charge that will likely remain relatively stable for long periods of time in the soil. However, the metals will be only weakly (physically) adsorbed to the biochar surface and immobilization may be easily reversed. The hypothesized trade-off between strength of sorption and longevity needs to be tested in a long term field trial before widespread application of biochar for metal contaminant remediation.

2.2.4 Precipitation with Mineral Ash

The source materials used for biochar production are rarely 100% organic in nature and inevitably contain minerals which, upon pyrolysis, remain entrained in the biochar matrix, contributing to the non-organic (or ash) portion of the resulting biochar. Mineral contents of source materials can range from <1% for woody biomass, up to ~25% for source materials derived from manure or crop residues. After high temperature pyrolysis the ash content in resulting biochars can be as high as 50% for manure-derived, or even 85% for bonemeal-derived, source materials (Amonette and Joseph 2009). Biochar ash is enriched in mineral salts of Na, K, Ca, Mg, P, S, Si and C, the concentrations of which increase with pyrolysis temperature (Gaskin et al. 2008). These elements are usually present in an oxidised form and therefore have surface properties that may complement metal immobilization. Uchimiya et al. (2010b) indicated the important contribution of lead phosphate precipitates in immobilising Pb in a broiler litter-derived biochar, as mentioned in section 2.1. Precipitation of Pb with phosphates contributed up to 87% of total Pb sorption in a dairy manure-derived biochar (Cao et al. 2009). Several different Pb phosphate minerals contributing to sorption in biochars have been identified, including hydrocerussite and hydroxypyromorphite (Cao et al. 2011), lead phosphate, and lead hydroxyapatite (Chen et al. 2006). Because these lead phosphate minerals have a very low solubility, their formation may be the reason for the capability of biochars to adsorb higher concentrations of Pb, compared to other divalent cations (Namgay et al. 2010, Uchimiya et al. 2010b, Trakal et al. 2011). Precipitation may also occur with other metals such as Cu, Cd, or Zn, as they also precipitate as insoluble phosphate and carbonate salts, mainly at high pH (Lindsay 1979).

2.2.5 Improvement of Soil Properties

The addition of biochar to soil leads to a number of changes in its physical, biological and chemical properties. These include increases in the soil microbial biomass, organic carbon, water holding capacity and nutrient use efficiency (Chan and Xu 2009, Major et al. 2009, Atkinson et al. 2010, Sohi et al. 2010, Lehmann et al. 2011). The results of these 'positive' effects on soil structure and function are that (i) the soil holds more water, (ii) erosion and runoff are reduced, (iii) leaching is prevented, and (iv) plant growth increases. The manipulation of the soil hydrological properties reduces the risk of metals in soils from entering surface- and ground-waters by leaching or erosion. Vegetation reduces wind and water erosion of soils as plant roots bind the soil and shoots act as a wind break (Tordoff et al. 2000, Mendez and Maier 2008). The development, and maintenance, of a vegetative cover is beneficial for both risk management and the wider, 'soft' aspects of ecological restoration (Dickinson et al. 2009). Therefore, the benefits of amending contaminated soils with biochars are not limited to the reduction in contaminant bioavailability. Biochar can deliver multiple benefits to risk management as part of an integrated remediation and ecological restoration project.

2.3 Possible Complications

Throughout this chapter and in previous studies and reviews, there are abundant examples of the beneficial effects of biochar application in soils; improving soil physical (drainage and aeration), chemical (supplying nutrients and water) and biological (suitable niche for soil microbiota) characteristics (Sohi et al. 2010). However, the application of biochar to contaminated soils should be implemented with a full understanding of the biogeochemical implications since amendments may result in the disturbance of natural biogeochemical cycles and, in some cases, increase environmental risk (Beesley et al. 2011).

The application of biochar to soil raises the pH and organic matter content (Hartley et al. 2009, Beesley et al. 2010b), increases the soil water holding capacity (Warnock et al. 2007) and may change redox conditions (a mixture of air-filled macropores and water-filled micropores may create microhabitats with oxic and anoxic conditions respectively). All of these changes may result in positive outcomes in contaminated soils, but may also increase the mobility and bioavailability of some contaminants. This is of particular concern in complex scenarios where multiple contaminants are present and biochar applications may have 'positive' effects (immobilization) for most of the contaminants but 'negative' effects (mobilization) for others.

This principle is illustrated by the the example of As, a metalloid frequently found in soils along with metals such as Cd, Cu, Pb or Zn (Conesa et al. 2008, Beesley et al. 2010a, Moreno-Jiménez et al. 2011). Arsenic is a trace element generally present in pristine soils and waters at low concentrations (e.g., <40 mg kg^{-1} in soils), but at anthropogenically contaminated sites it can be present at elevated concentrations in the soil and thereafter transferred to surface and ground waters or into ecosystems (Moreno-Jiménez et al. 2012). Arsenic is toxic, even at low concentrations (World Health Organization standard for drinking water$=$ 10 µg L^{-1}) since it is carcinogenic (Williams et al. 2005) and can disturb cell metabolism and homeostasis (Meharg and Hartley-Whitaker 2002, Patra et al. 2004). Arsenic, unlike metals, is not present in soils as a cation but as an oxy-anion. Therefore, it has a different geochemical fate (Adriano 2001). Even when biochar is able to retain cationic metals, its addition to multi-element contaminated soils can lead to the mobilization of As (Hartley et al. 2009, Beesley et al. 2010b). The main indirect effects that may lead to As mobilization in amended soils are (i) an increase in soil pH, (ii) an increase in DOC, (iii) a reduction in redox potential, and (iv) greater phosphate mobility.

2.3.1 Increase in Soil pH

Arsenic binds to positively charged surfaces such as Fe and Mn oxides in soils, as opposed to cationic metals which are bound to the negatively charged surfaces such as clay minerals and organic matter. When the soil pH is increased, metals are increasingly bound to these negatively charged surfaces. Contrary to cationic metals, As is released from positively charged soil surfaces when the soil pH is increased. Several studies (discussed in section 2.1) have reported that soil pH increases after biochar application to contaminated soils (Beesley et al. 2010b, Beesley et al. 2011), and an increase of soil pH has been reported to increase As mobility and uptake by organisms (Fitz and Wenzel 2002, Moreno-Jiménez et al. 2012). Therefore, biochar addition may increase the pH of a soil and increase the bioavailability of As.

2.3.2 Application of Organic Matter

Biochar application to soils increases stable organic matter (Cheng et al. 2008, Bell and Worrall 2011) and may either increase (Hartley et al. 2009, Novak et al. 2009b, Beesley et al. 2010b) or decrease DOC by providing binding sites for organic matter to sorb to (Gomez-Eyles et al. 2011). Arsenic is less attracted to negatively charged biochar surfaces than metals due to its oxy-anionic behavior (Fitz and Wenzel 2002, Beesley and Marmiroli 2011). Therefore As is not immobilized by biochar to the same extent as cationic metals. When biochars increase DOC it is likely that As mobility will also increase, as these two parameters are usually positively correlated to each other. The underlying mechanism for this co-mobilization is unclear, although Mikutta and Kretzschmar (2011) observed a ternary complex formation between arsenate and ferric iron complexes of humic substances which could be responsible for this effect. It is also possible for the DOC to compete with arsenic for retention sites on soil surfaces (Fitz and Wenzel 2002), also resulting in high concentrations of soluble As with high concentrations of DOC (Hartley et al. 2009). This co-mobilization of As is likely to reduce biochar effectiveness for soil remediation purposes (Mench et al. 2003, Bernal et al. 2007, Moreno-Jiménez et al. 2012). Conversely, As is methylated in soil in the presence of (non-charred) organic matter (Oremland and Stolz 2003), and methylated species of As are less toxic than inorganic As (Hughes 2002). This means that due to the application of biochars which increase labile carbon, As speciation may be changed to a less toxic form. However, to date, the methylation of As induced by biochar application in soil has not been reported.

2.3.3 Changes in Redox Conditions

The application of biochar to soil results in an increase in soil porosity and a decrease in bulk density (Warnock et al. 2007, Atkinson et al. 2010). This change in soil hydrology leads to large air-filled (oxic) pores and some small water filled pores where anoxic conditions may persist. Arsenic speciation is sensitive to changes in redox conditions. In anoxic conditions As will predominantly be in the reduced form, As(III), which is more mobile and toxic than As(V), found in oxic conditions (Ratnaike 2003, Moreno-Jiménez et al. 2012). Although generally a decrease in soil bulk density increases aeration, small micropores in biochar may have locally reducing conditions. These redox microhabitats may result in a reduction in the valence state of As. If these microhabitats are in contact with biota (i.e., plant root hairs), organisms may be exposed to elevated levels of As(III), increasing uptake and potentially enhancing toxic effects. However, the effect of biochar on the redox potential of meso and microscale pores in soils is currently a gap in the literature.

2.3.4 Phosphorus Supply

Contaminated soils are often poor in nutrients and fertilizer application is required for plant growth. Some biochars (such as those derived from dairy and poultry litter manures) are good sources of bioavailable phosphorus (Sohi et al. 2010). Phosphate is chemically very similar to As(V), so it can displace As(V) from soil surfaces, resulting in As release into solution. Phosphate and As(V) are also analogues in plant uptake (Meharg and Macnair 1992), so plants can be protected against As(V) uptake by high concentrations of soluble P (Moreno-Jiménez et al. 2012). However, the risk of As leaching into surface and ground waters will be increased if the soluble fraction of As is not taken up by plants (Fitz and Wenzel 2002), and there is also the risk of these poultry litter chars becoming a source of phosphate nutrient contamination.

2.3.5 Summary of Complications

All of the aforementioned issues make the application of biochar into soils containing elevated As a complex and contentious practice, so further research is needed to fully understand all the underlying mechanisms. Several studies have reported an increase in toxicity due to the interaction of As and biochar during remediation applications (Hartley et al. 2009, Beesley et al. 2010b, Beesley et al. 2011) but this may be corrected in some cases by co-applying an effective As sorbent, e.g., iron oxides as described in section 2.4. As well as arsenic, other elements are more mobile with increasing pH (e.g., Sb, W, Mo) (Alloway 1995), are sensitive to redox changes (e.g., Cr) (Kozuh et al. 1999), co-mobilized with DOC (e.g., Cu, Pb and Hg) (Drexel et al. 2002, Bernal et al. 2007), or display antagonistic or synergistic behavior during soil retention and plant uptake (e.g., Cd/Zn) (Smilde et al. 1992). These factors should also be taken into account on a site specific basis when considering biochar amendments for remediation.

The potential transfer of contaminants to the food chain should be addressed when a soil is amended with biochar. Organic soil amendments usually promote growth and, ultimately, increase the biomass of soil organisms (plants, invertebrates, etc.). These organisms can be a primary source of trace elements into the terrestrial food chain, inducing problems to secondary consumers and humans (when biochar is used in agriculture). Karami et al. (2011) highlighted that studies of contaminated soil remediation usually just evaluate the bioavailability of a contaminant as a function of reductions in mobility achieved by the amendment. In some cases we therefore fail to take into account of the introduction of a new pathway as the introduction of an amendment may increase the amount of metal entering the food chain by increasing the plant uptake and biomass.

2.4 Impregnating Biochars with Iron Oxides

In a multi-element contaminated soil context, the co-application of organic and inorganic amendments can produce better results than applying the amendments individually (Mench et al. 2003). Iron oxides are known as an effective sink for trace elements such as As, Hg, Se, Cr, Pb, etc., so they have classically been used for soil remediation of inorganic pollutants (Warren et al. 2003, Waychunas et al. 2005). Other oxides, such as Al and Mn oxides show similar properties for immobilization. They have a high affinity for some contaminants and iron oxide phases can retain metals and arsenic due to the formation of chemical bonds with surface atoms (chemosorption), forming covalent, ionic or hydrogen bonds by inner and outer sphere complexation (Waychunas et al. 2005). Iron-oxides particularly, have attracted attention as they are able to effectively immobilize As (Dixit and Hering 2003). The effects of iron oxides (e.g., goethite, ferrihydrite) in contaminated soils have been studied by applying them straight into the soil in a variety of forms, namely, red mud, Fe salts, Fe(0), or iron-enriched by-products. In all cases, adding iron oxides to soil increased the immobilization of trace elements (Hartley et al. 2004, Friesl et al. 2006, Kumpiene et al. 2008, Vangronsveld et al. 2009). It may therefore be possible to enhance the metal pollutant binding capacity of biochars by impregnating them with iron oxides. This would be especially beneficial for reducing As mobilization, which is discussed in section 2.3 as a possible complication arising from amending contaminated soils with biochar. The use of iron oxide impregnated sorbents has been studied for the water treatment industry. Activated carbons have been impregnated with iron oxides to combine the high surface area provided by the AC with the enhanced sorptive properties of the iron oxides (Reed 2000, Vaughan and Reed 2005). Biochars and ACs impregnated with iron-oxides can be produced by soaking the precursor material (e.g., citrus peel or pinewood) in an iron chloride solution before pyrolysis to incorporate the iron oxides into the biochar structure during pyrolysis (Chen et al. 2011a). Other authors have soaked the biochar itself in an iron solution and then used a solution of NaOH to induce the formation of amorphous Fe-(hydro)oxides in the charred material by basification (Muñiz et al. 2009, Chang et al. 2010). The removal of As from solution by the impregnated AC was 1–2 orders of magnitude higher than by the virgin AC (Reed 2000). The economic cost of producing these 'speciality' biochars may make them unsuitable for large scale soil remediation projects, but they may be viable on a site specific basis.

Biochar can be also impregnated with inorganic compounds other than iron oxides, such as phosphates or carbonates, which are able to enhance metal immobilization, but these may also mobilize As, so caution is required before applying these to polluted soils.

Conclusion

As is discussed throughout this book and in the scientific literature, biochars have properties that enable them to enhance soil quality, fertility and structure (Chan et al. 2007, Novak et al. 2009a, Thies and Rillig 2009), and sequester carbon (Lehmann et al. 2006, Kuzyakov et al. 2009, McHenry 2009). We have shown here that there also is a growing body of evidence supporting the application of biochars to contaminated soils. They can be suitable for the remediation of soils contaminated with both organic and inorganic contaminants. Although biochars of certain properties are better suited for certain types of contaminants, a good general understanding of biochar sorption mechanisms can enable the selection of the most suitable biochar for the contaminated site in question. However, there are a number of caveats and complications arising from biochar soil amendments, especially at multi-element contaminated sites. It is therefore important to consider the site biogeochemistry before biochars are applied for remediation purposes.

References

Abrahams, P.W. 2002. Soils: their implications to human health. Sci. Total Environ. 291: 1–32.

Accardi-Dey, A. and P.M. Gschwend. 2001. Assessing the Combined Roles of Natural Organic Matter and Black Carbon as Sorbents in Sediments. Environ. Sci. Technol. 36: 21–29.

Adriano, D.C. 2001. Trace elements in terrestrial environments: biogeochemistry, bioavailability, and risks of metals. Springer-Verlag, New York. USA.

Ahmad, M., S. Soo Lee, J.E. Yang, H.-M. Ro, Y. Han Lee, and Y. Sik Ok. 2012. Effects of soil dilution and amendments (mussel shell, cow bone, and biochar) on Pb availability and phytotoxicity in military shooting range soil. Ecotoxicol. Environ. Saf. 79: 225–231.

Al-Qaessi, F. and L. Abu-Farah. 2010. Activated Carbon Production from Date Stones Using Phosphoric Acid. Energy Sources, Part A: Recovery, Utilization, and Environmental Effects 32: 1316–1325.

Alloway, B.J. 1995. Heavy metals in soils. Blackie Academic & Professional, Glasgow, UK.

Amonette, J. and S. Joseph. 2009. Characteristics of biochar: microchemical properties. Biochar for Environmental Management: Science and Technology 33.

Atkinson, C., J. Fitzgerald, and N. Hipps. 2010. Potential mechanisms for achieving agricultural benefits from biochar application to temperate soils: a review. Plant Soil 337: 1–18.

Azargohar, R. and A.K. Dalai. 2008. Steam and KOH activation of biochar: Experimental and modeling studies. Microporous and Mesoporous Materials 110: 413–421.

Beckingham, B. and U. Ghosh. 2011. Field-Scale Reduction of PCB Bioavailability with Activated Carbon Amendment to River Sediments. Environ. Sci. Technol. 45: 10567–10574.

Beesley, L. and N. Dickinson. 2010. Carbon and trace element fluxes in the pore water of an urban soil following greenwaste compost, woody and biochar amendments, inoculated with the earthworm Lumbricus terrestris. Soil Biol. Biochem. 43: 188–196.

Beesley, L. and M. Marmiroli. 2011. The immobilisation and retention of soluble arsenic, cadmium and zinc by biochar. Environ. Pollut. 159: 474–480.

Beesley, L., E. Moreno-Jimenez, R. Clemente, N. Lepp, and N. Dickinson. 2010a. Mobility of arsenic, cadmium and zinc in a multi-element contaminated soil profile assessed by *in situ* soil pore water sampling, column leaching and sequential extraction. Environ. Pollut. 158: 155–160.

Beesley, L., E. Moreno-Jiménez, and J.L. Gomez-Eyles. 2010b. Effects of biochar and greenwaste compost amendments on mobility, bioavailability and toxicity of inorganic and organic contaminants in a multi-element polluted soil. Environ. Pollut. 158: 2282–2287.

Beesley, L., E. Moreno-Jiménez, J.L. Gomez-Eyles, E. Harris, B. Robinson, and T. Sizmur. 2011. A review of biochars' potential role in the remediation, revegetation and restoration of contaminated soils. Environ. Pollut. 159: 3269–3282.

Bell, M.J. and F. Worrall. 2011. Charcoal addition to soils in N.E. England: A carbon sink with environmental co-benefits? Sci. Total Environ. 409: 1704–1714.

Bernal, M., R. Clemente, and D. Walker. The role of organic amendment in the bioremediation of heavy metal-polluted soils. pp. 2–58. *In*: R.W. Gore (ed.). 2007. Environmental Research at the Leading Edge. Nova Publishers, New York, USA.

Bolan, N.S. and V. Duraisamy. 2003. Role of inorganic and organic soil amendments on immobilisation and phytoavailability of heavy metals: a review involving specific case studies. Aust. J. Soil Res. 41: 533–555.

Brändli, R.C., T. Hartnik, T. Henriksen, and G. Cornelissen. 2008. Sorption of native polyaromatic hydrocarbons (PAH) to black carbon and amended activated carbon in soil. Chemosphere 73: 1805–1810.

Bucheli, T.D. and O. Gustafsson. 2000. Quantification of the soot-water distribution coefficient of PAHs provides mechanistic basis for enhanced sorption observations. Environ. Sci. Technol. 34: 5144–5151.

Bushnaf, K.M., S. Puricelli, S. Saponaro, and D. Werner. 2011. Effect of biochar on the fate of volatile petroleum hydrocarbons in an aerobic sandy soil. Journal of contaminant hydrology 126: 208–215.

Cao, X. and W. Harris. 2010. Properties of dairy-manure-derived biochar pertinent to its potential use in remediation. Bioresour. Technol. 101: 5222–5228.

Cao, X., L. Ma, Y. Liang, B. Gao, and W. Harris. 2011. Simultaneous Immobilization of Lead and Atrazine in Contaminated Soils Using Dairy-Manure Biochar. Environ. Sci. Technol. 45: 4884–4889.

Cao, X.D., L.N. Ma, B. Gao, and W. Harris. 2009. Dairy-Manure Derived Biochar Effectively Sorbs Lead and Atrazine. Environ. Sci. Technol. 43: 3285–3291.

Chai, Y., R.J. Currie, J.W. Davis, M. Wilken, G.D. Martin, V.N. Fishman, and U. Ghosh. 2012. Effectiveness of Activated Carbon and Biochar in Reducing the Availability of Polychlorinated Dibenzo-p-dioxins/Dibenzofurans in Soils. Environ. Sci. Technol. 46: 1035–1043.

Chan, K., L. Van Zwieten, I. Meszaros, A. Downie, and S. Joseph. 2007. Agronomic values of greenwaste biochar as a soil amendment. Aust. J. Soil Res. 45: 629–634.

Chan, K. and Z. Xu. Biochar: nutrient properties and their enhancement. p. 67. *In*: J. Lehmann, S. Joseph (eds.). 2009. Biochar for Environmental Management: Science and Technology. Earthscan.

Chang, Q., W. Lin, and W.-c. Ying. 2010. Preparation of iron-impregnated granular activated carbon for arsenic removal from drinking water. J. Hazard. Mater. 184: 515–522.

Chen, B. and W. Huang. 2011. Effects of compositional heterogeneity and nanoporosity of raw and treated biomass-generated soot on adsorption and absorption of organic contaminants. Environ. Pollut. 159: 550–556.

Chen, B.L., Z.M. Chen, and S.F. Lv. 2011a. A novel magnetic biochar efficiently sorbs organic pollutants and phosphate. Bioresour. Technol. 102: 716–723.

Chen, B.L., D.D. Zhou, and L.Z. Zhu. 2008. Transitional adsorption and partition of nonpolar and polar aromatic contaminants by biochars of pine needles with different pyrolytic temperatures. Environ. Sci. Technol. 42: 5137–5143.

Chen, S.-B., Y.-G. Zhu, Y.-B. Ma, and G. McKay. 2006. Effect of bone char application on Pb bioavailability in a Pb-contaminated soil. Environ. Pollut. 139: 433–439.

Chen, X., G. Chen, L. Chen, Y. Chen, J. Lehmann, M.B. McBride, and A.G. Hay. 2011b. Adsorption of copper and zinc by biochars produced from pyrolysis of hardwood and corn straw in aqueous solution. Bioresour. Technol. 102: 8877–8884.

Cheng, C.H., J. Lehmann, and M.H. Engelhard. 2008. Natural oxidation of black carbon in soils: Changes in molecular form and surface charge along a climosequence. Geochim. Cosmochim. Acta 72: 1598–1610.

Cheng, C.H., J. Lehmann, J.E. Thies, S.D. Burton, and M.H. Engelhard. 2006. Oxidation of black carbon by biotic and abiotic processes. Org. Geochem. 37: 1477–1488.

Cho, Y.M., U. Ghosh, A.J. Kennedy, A. Grossman, G. Ray, J.E. Tomaszewski, D.W. Smithenry, T.S. Bridges, and R.G. Luthy. 2009. Field Application of Activated Carbon Amendment for *in situ* Stabilization of Polychlorinated Biphenyls in Marine Sediment. Environ. Sci. Technol. 43: 3815–3823.

Cho, Y.M., D.W. Smithenry, U. Ghosh, A.J. Kennedy, R.N. Millward, T.S. Bridges, and R.G. Luthy. 2007. Field methods for amending marine sediment with activated carbon and assessing treatment effectiveness. Mar. Environ. Res. 64: 541–555.

Chun, Y., G. Sheng, C.T. Chiou, and B. Xing. 2004. Compositions and Sorptive Properties of Crop Residue-Derived Chars. Environ. Sci. Technol. 38: 4649–4655.

Conesa, H.M., B.H. Robinson, R. Schulin, and B. Nowack. 2008. Metal extractability in acidic and neutral mine tailings from the Cartagena-La Unión Mining District (SE Spain). Appl. Geochem. 23: 1232–1240.

Connell, D.W. 2005. Basic concepts of environmental chemistry. CRC/Taylor & Francis.

Cornelissen, G., G.D. Breedveld, S. Kalaitzidis, K. Christanis, A. Kibsgaard, and A.M.P. Oen. 2006a. Strong sorption of native PAHs to pyrogenic and unburned carbonaceous geosorbents in sediments. Environ. Sci. Technol. 40: 1197–1203.

Cornelissen, G., G.D. Breedveld, K. Naes, A.M.P. Oen, and A. Ruus. 2006b. Bioaccumulation of native polycyclic aromatic hydrocarbons from sediment by a polychaete and a gastropod: Freely dissolved concentrations and activated carbon amendment. Environ. Toxicol. Chem. 25: 2349–2355.

Cornelissen, G. and O. Gustafsson. 2005. Importance of unburned coal carbon, black carbon, and amorphous organic carbon to phenanthrene sorption in sediments. Environ. Sci. Technol. 39: 764–769.

Cornelissen, G., O. Gustafsson, T.D. Bucheli, M.T.O. Jonker, A.A. Koelmans, and P.C.M. Van Noort. 2005. Extensive sorption of organic compounds to black carbon, coal, and kerogen in sediments and soils: Mechanisms and consequences for distribution, bioaccumulation, and biodegradation. Environ. Sci. Technol. 39: 6881–6895.

Dickinson, N.M., A.J.M. Baker, A. Doronila, S. Laidlaw, and R.D. Reeves. 2009. Phytoremediation of inorganics: realism and synergies. Int. J. Phytoremediation 11: 97–114.

Dixit, S. and J.G. Hering. 2003. Comparison of Arsenic(V) and Arsenic(III) Sorption onto Iron Oxide Minerals: Implications for Arsenic Mobility. Environ. Sci. Technol. 37: 4182–4189.

Drexel, R.T., M. Haitzer, J.N. Ryan, G.R. Aiken, and K.L. Nagy. 2002. Mercury(II) Sorption to Two Florida Everglades Peats: Evidence for Strong and Weak Binding and Competition by Dissolved Organic Matter Released from the Peat. Environ. Sci. Technol. 36: 4058–4064.

Fagervold, S.K., Y.Z. Chai, J.W. Davis, M. Wilken, G. Cornelissen, and U. Ghosh. 2010. Bioaccumulation of Polychlorinated Dibenzo-p-Dioxins/Dibenzofurans in E. fetida from Floodplain Soils and the Effect of Activated Carbon Amendment. Environ. Sci. Technol. 44: 5546–5552.

Fellet, G., L. Marchiol, G. Delle Vedove, and A. Peressotti. 2011. Application of biochar on mine tailings: Effects and perspectives for land reclamation. Chemosphere 83: 1262–1267.

Fitz, W.J. and W.W. Wenzel. 2002. Arsenic transformations in the soil-rhizosphere-plant system: fundamentals and potential application to phytoremediation. J. Biotechnol. 99: 259–278.

Friesl, W., J. Friedl, K. Platzer, O. Horak, and M.H. Gerzabek. 2006. Remediation of contaminated agricultural soils near a former Pb/Zn smelter in Austria: Batch, pot and field experiments. Environ. Pollut. 144: 40–50.

Gaskin, J., C. Steiner, K. Harris, K. Das, and B. Bibens. 2008. Effect of low-temperature pyrolysis conditions on biochar for agricultural use. Transactions of the American Society of Agricultural and Biological Engineers 51: 2061–2069.

Gell, K., J. van Groenigen, and M.L. Cayuela. 2011. Residues of bioenergy production chains as soil amendments: Immediate and temporal phytotoxicity. J. Hazard. Mater. 186: 2017–2025.

Ghosh, U. 2007. The role of black carbon in influencing availability of PAHs in sediments. Hum. Ecol. Risk Assess. 13: 276–285.

Ghosh, U., J.S. Gillette, R.G. Luthy, and R.N. Zare. 2000. Microscale Location, Characterization, and Association of Polycyclic Aromatic Hydrocarbons on Harbor Sediment Particles. Environ. Sci. Technol. 34: 1729–1736.

Ghosh, U., R.G. Luthy, G. Cornelissen, D. Werner, and C.A. Menzie. 2011. *In situ* Sorbent Amendments: A New Direction in Contaminated Sediment Management. Environ. Sci. Technol. 45: 1163–1168.

Ghosh, U., J.R. Zimmerman, and R.G. Luthy. 2003. PCB and PAH speciation among particle types in contaminated harbor sediments and effects on PAH bioavailability. Environ. Sci. Technol. 37: 2209–2217.

Gomez-Eyles, J.L., M.T.O. Jonker, M.E. Hodson, and C.D. Collins. 2012. Passive Samplers Provide a Better Prediction of PAH Bioaccumulation in Earthworms and Plant Roots than Exhaustive, Mild Solvent, and Cyclodextrin Extractions. Environ. Sci. Technol. 46: 962–969.

Gomez-Eyles, J.L., T. Sizmur, C.D. Collins, and M.E. Hodson. 2011. Effects of biochar and the earthworm Eisenia fetida on the bioavailability of polycyclic aromatic hydrocarbons and potentially toxic elements. Environ. Pollut. 159: 616–622.

Gundale, M. and T. DeLuca. 2007. Charcoal effects on soil solution chemistry and growth of *Koeleria macrantha* in the ponderosa pine/Douglas-fir ecosystem. Biol. Fertility Soils 43: 303–311.

Hale, S., K. Hanley, J. Lehmann, A. Zimmerman, and G. Cornelissen. 2011. Effects of Chemical, Biological, and Physical Aging As Well As Soil Addition on the Sorption of Pyrene to Activated Carbon and Biochar. Environ. Sci. Technol. 45: 10445–10453.

Hale, S.E. and D. Werner. 2010. Modeling the Mass Transfer of Hydrophobic Organic Pollutants in Briefly and Continuously Mixed Sediment after Amendment with Activated Carbon. Environ. Sci. Technol. 44: 3381–3387.

Hartley, W., N.M. Dickinson, P. Riby, and N.W. Lepp. 2009. Arsenic mobility in brownfield soils amended with green waste compost or biochar and planted with Miscanthus. Environ. Pollut. 157: 2654–2662.

Hartley, W., R. Edwards, and N.W. Lepp. 2004. Arsenic and heavy metal mobility in iron oxide-amended contaminated soils as evaluated by short- and long-term leaching tests. Environ. Pollut. 131: 495–504.

Harvey, O.R., B.E. Herbert, R.D. Rhue, and L.-J. Kuo. 2011. Metal Interactions at the Biochar-Water Interface: Energetics and Structure-Sorption Relationships Elucidated by Flow Adsorption Microcalorimetry. Environ. Sci. Technol. 45: 5550–5556.

Hughes, M.F. 2002. Arsenic toxicity and potential mechanisms of action. Toxicol. Lett. 133: 1–16.

Jaramillo, J., V. Gomez-Serrano, and P.M. Alvarez. 2009. Enhanced adsorption of metal ions onto functionalized granular activated carbons prepared from cherry stones. J. Hazard. Mater. 161: 670–676.

Jonker, M.T.O. and A.A. Koelmans. 2002. Sorption of Polycyclic Aromatic Hydrocarbons and Polychlorinated Biphenyls to Soot and Soot-like Materials in the Aqueous Environment: Mechanistic Considerations. Environ. Sci. Technol. 36: 3725–3734.

Jonker, M.T.O., M.P.W. Suijkerbuijk, H. Schmitt, and T.L. Sinnige. 2009. Ecotoxicological Effects of Activated Carbon Addition to Sediments. Environ. Sci. Technol. 43: 5959–5966.

Jonker, M.T.O., S.A. van der Heijden, J.P. Kreitinger, and S.B. Hawthorne. 2007. Predicting PAH bioaccumulation and toxicity in earthworms exposed to manufactured gas plant soils with solid-phase microextraction. Environ. Sci. Technol. 41: 7472–7478.

Kannan, N. and G. Rengasamy. 2005. Comparison of Cadmium Ion Adsorption on Various ACTIVATED CARBONS. Water, Air, Soil Pollut. 163: 185–201.

Karami, N., R. Clemente, E. Moreno-Jiménez, N.W. Lepp, and L. Beesley. 2011. Efficiency of green waste compost and biochar soil amendments for reducing lead and copper mobility and uptake to ryegrass. J. Hazard. Mater. 191: 41–48.

Kasozi, G.N., A.R. Zimmerman, P. Nkedi-Kizza, and B. Gao. 2010. Catechol and Humic Acid Sorption onto a Range of Laboratory-Produced Black Carbons (Biochars). Environ. Sci. Technol. 44: 6189–6195.

Kookana, R.S. 2010. The role of biochar in modifying the environmental fate, bioavailability, and efficacy of pesticides in soils: a review. Aust. J. Soil Res. 48: 627–637.

Kozuh, N., J. Stupar, and B. Gorenc. 1999. Reduction and Oxidation Processes of Chromium in Soils. Environ. Sci. Technol. 34: 112–119.

Kumpiene, J., A. Lagerkvist, and C. Maurice. 2008. Stabilization of As, Cr, Cu, Pb and Zn in soil using amendments—A review. Waste Manage. 28: 215–225.

Kuzyakov, Y., I. Subbotina, H. Chen, I. Bogomolova, and X. Xu. 2009. Black carbon decomposition and incorporation into soil microbial biomass estimated by ^{14}C labeling. Soil Biol. Biochem. 41: 210–219.

Laird, D., P. Fleming, B. Wang, R. Horton, and D. Karlen. 2010. Biochar impact on nutrient leaching from a Midwestern agricultural soil. Geoderma 158: 436–442.

Lee, J.W., M. Kidder, B.R. Evans, S. Paik, A.C. Buchanan III, C.T. Garten, and R.C. Brown. 2010. Characterization of Biochars Produced from Cornstovers for Soil Amendment. Environ. Sci. Technol. 44: 7970–7974.

Leglize, P., S. Alain, B. Jacques, and L. Corinne. 2008. Adsorption of phenanthrene on activated carbon increases mineralization rate by specific bacteria. J. Hazard. Mater. 151: 339–347.

Lehmann, J., J. Gaunt, and M. Rondon. 2006. Biochar Sequestration in Terrestrial Ecosystems—A Review. Mitigation and Adaptation Strategies for Global Change 11: 395–419.

Lehmann, J., M.C. Rillig, J. Thies, C.A. Masiello, W.C. Hockaday, and D. Crowley. 2011. Biochar effects on soil biota—A review. Soil Biol. Biochem. 43: 1812–1836.

Liang, B., J. Lehmann, D. Solomon, J. Kinyangi, J. Grossman, B. O'Neill, J.O. Skjemstad, J. Thies, F.J. Luizao, J. Petersen, and E.G. Neves. 2006. Black carbon increases cation exchange capacity in soils. Soil Sci. Soc. Am. J. 70: 1719–1730.

Liesch, A., S. Weyers, J. Gaskin, and K. Das. 2010. Impact of two different biochars on earthworm growth and survival. Annals of Environmental Science 4: 1.

Lim, W.C., C. Srinivasakannan, and N. Balasubramanian. 2010. Activation of palm shells by phosphoric acid impregnation for high yielding activated carbon. J. Anal. Appl. Pyrolysis 88: 181–186.

Liu, Z. and F.-S. Zhang. 2009. Removal of lead from water using biochars prepared from hydrothermal liquefaction of biomass. J. Hazard. Mater. 167: 933–939.

Major, J., C. Steiner, A. Downie, and J. Lehmann. Biochar effects on nutrient leaching. pp. 273–284. *In*: J. Lehmann, S. Joseph (eds.). 2009. Biochar for Environmental Management: Science and Technology. Earthscan.

McBeath, A.V., R.J. Smernik, M.P.W. Schneider, M.W.I. Schmidt, and E.L. Plant. 2011. Determination of the aromaticity and the degree of aromatic condensation of a thermosequence of wood charcoal using NMR. Org. Geochem. 42: 1194–1202.

McHenry, M.P. 2009. Agricultural bio-char production, renewable energy generation and farm carbon sequestration in Western Australia: Certainty, uncertainty and risk. Agric., Ecosyst. Environ. 129: 1–7.

Meharg, A.A. and J. Hartley-Whitaker. 2002. Arsenic uptake and metabolism in arsenic resistant and nonresistant plant species. New Phytol. 154: 29–43.

Meharg, A.A. and M.R. Macnair. 1992. Suppression of the High Affinity Phosphate Uptake System: A Mechanism of Arsenate Tolerance in Holcus lanatus L. J. Exp. Bot. 43: 519–524.

Mench, M., S. Bussière, J. Boisson, E. Castaing, J. Vangronsveld, A. Ruttens, T. De Koe, P. Bleeker, A. Assunção, and A. Manceau. 2003. Progress in remediation and revegetation of the barren Jales gold mine spoil after *in situ* treatments. Plant Soil 249: 187–202.

Mench, M., N. Lepp, V. Bert, J.-P. Schwitzguébel, S. Gawronski, P. Schröder, and J. Vangronsveld. 2010. Successes and limitations of phytotechnologies at field scale: outcomes, assessment and outlook from COST Action 859. J. Soils Sediments 10: 1039–1070.

Mendez, M. and R. Maier. 2008. Phytostabilization of mine tailings in arid and semiarid environments—an emerging remediation technology. Environ. Health Perspect. 116: 278.

Mikutta, C. and R. Kretzschmar. 2011. Spectroscopic Evidence for Ternary Complex Formation between Arsenate and Ferric Iron Complexes of Humic Substances. Environ. Sci. Technol. 45: 9550–9557.

Millward, R.N., T.S. Bridges, U. Ghosh, J.R. Zimmerman, and R.G. Luthy. 2005. Addition of activated carbon to sediments to reduce PCB bioaccumulation by a polychaete (Neanthes arenaceodentata) and an amphipod (Leptocheirus plumulosus). Environ. Sci. Technol. 39: 2880–2887.

Mohan, D., R. Sharma, V.K. Singh, P. Steele, and C.U. Pittman. 2012. Fluoride Removal from Water using Bio-Char, a Green Waste, Low-Cost Adsorbent: Equilibrium Uptake and Sorption Dynamics Modeling. Industrial & Engineering Chemistry Research 51: 900–914.

Moreno-Jiménez, E., E. Esteban, and J.M. Peñalosa. The Fate of Arsenic in Soil-Plant Systems Reviews of Environmental Contamination and Toxicology. pp. 1–37. *In*: D.M. Whitacre. (ed.). 2012. Springer New York.

Moreno-Jiménez, E., S. Vázquez, R.O. Carpena-Ruiz, E. Esteban, and J.M. Peñalosa. 2011. Using Mediterranean shrubs for the phytoremediation of a soil impacted by pyritic wastes in Southern Spain: A field experiment. J. Environ. Manage 92: 1584–1590.

Mukherjee, A., A.R. Zimmerman, and W. Harris. 2011. Surface chemistry variations among a series of laboratory-produced biochars. Geoderma 163: 247–255.

Muñiz, G., V. Fierro, A. Celzard, G. Furdin, G. Gonzalez-Sanchez, and M.L. Ballinas. 2009. Synthesis, characterization and performance in arsenic removal of iron-doped activated carbons prepared by impregnation with Fe(III) and Fe(II). J. Hazard. Mater. 165: 893–902.

Nag, S.K., R. Kookana, L. Smith, E. Krull, L.M. Macdonald, and G. Gill. 2011. Poor efficacy of herbicides in biochar-amended soils as affected by their chemistry and mode of action. Chemosphere 84: 1572–1577.

Namgay, T., B. Singh, and B.P. Singh. 2010. Influence of biochar application to soil on the availability of As, Cd, Cu, Pb, and Zn to maize (Zea mays L.). Aust. J. Soil Res. 48: 638–647.

Novak, J.M., W.J. Busscher, D.L. Laird, M. Ahmedna, D.W. Watts, and M.A.S. Niandou. 2009a. Impact of Biochar Amendment on Fertility of a Southeastern Coastal Plain Soil. Soil Sci. 174: 105–112.

Novak, J.M., W.J. Busscher, D.W. Watts, D.A. Laird, M.A. Ahmedna, and M.A.S. Niandou. 2009b. Short-term CO2 mineralization after additions of biochar and switchgrass to a Typic Kandiudult. Geoderma 154: 281–288.

Oremland, R.S. and J.F. Stolz. 2003. The Ecology of Arsenic. Science 300: 939–944.

Patra, M., N. Bhowmik, B. Bandopadhyay, and A. Sharma. 2004. Comparison of mercury, lead and arsenic with respect to genotoxic effects on plant systems and the development of genetic tolerance. Environ. Exp. Bot. 52: 199–223.

Paul, P. and U. Ghosh. 2011. Influence of activated carbon amendment on the accumulation and elimination of PCBs in the earthworm Eisenia fetida. Environ. Pollut. 159: 3763–3768.

Pellera, F.-M., A. Giannis, D. Kalderis, K. Anastasiadou, R. Stegmann, J.-Y. Wang, and E. Gidarakos. 2012. Adsorption of Cu(II) ions from aqueous solutions on biochars prepared from agricultural by-products. J. Environ. Manage. 96: 35–42.

Pietikäinen, J., O. Kiikkilä, and H. Fritze. 2000. Charcoal as a habitat for microbes and its effect on the microbial community of the underlying humus. Oikos 89: 231–242.

Pignatello, J.J. and B. Xing. 1995. Mechanisms of Slow Sorption of Organic Chemicals to Natural Particles. Environ. Sci. Technol. 30: 1–11.

Ratnaike, R.N. 2003. Acute and chronic arsenic toxicity. Postgrad. Med. J. 79: 391–396.

Reed, B. 2000. Adsorption of heavy metals using iron impregnated activated carbon. J. Environ. Eng. 126: 896–874.

Rhodes, A.H., A. Carlin, and K.T. Semple. 2008. Impact of black carbon in the extraction and mineralization of phenanthrene in soil. Environ. Sci. Technol. 42: 740–745.

Sanchez-Polo, M. and J. Rivera-Utrilla. 2002. Adsorbent-adsorbate interactions in the adsorption of Cd(II) and Hg(II) on ozonized activated carbons. Environ. Sci. Technol. 36: 3850–3854.

Shen, Y.-S., S.-L. Wang, Y.-M. Tzou, Y.-Y. Yan, and W.-H. Kuan. 2012. Removal of hexavalent Cr by coconut coir and derived chars—The effect of surface functionality. Bioresour. Technol. 104: 165–172.

Simoes, J.A.M. and J.L. Beauchamp. 1990. Transition metal-hydrogen and metal-carbon bond strengths: the keys to catalysis. Chemical Reviews 90: 629–688.

Simon, L. Effects of Natural Zeolite and Bentonite on the Phytoavailability of Heavy Metals in Chicory. pp. 261–271. 2000. Environmental Restoration of Metals-Contaminated Soils. CRC Press.

Sizmur, T., J. Wingate, T. Hutchings, and M.E. Hodson. 2011. Lumbricus terrestris L. does not impact on the remediation efficiency of compost and biochar amendments. Pedobiologia 54, Supplement: S211–S216.

Smilde, K.W., B. Luit, and W. Driel. 1992. The extraction by soil and absorption by plants of applied zinc and cadmium. Plant Soil 143: 233–238.

Sohi, S., E. Krull, E. Lopez-Capel, and R. Bol. 2010. A review of biochar and its use and function in soil. Advances in Agronomy: 47–82.

Spokas, K.A., W.C. Koskinen, J.M. Baker, and D.C. Reicosky. 2009. Impacts of woodchip biochar additions on greenhouse gas production and sorption/degradation of two herbicides in a Minnesota soil. Chemosphere 77: 574–581.

Sun, K., M. Keiluweit, M. Kleber, Z. Pan, and B. Xing. 2011. Sorption of fluorinated herbicides to plant biomass-derived biochars as a function of molecular structure. Bioresour. Technol. 102: 9897–9903.

Sun, X.L. and U. Ghosh. 2008. The effect of activated carbon on partitioning, desorption, and biouptake of native polychlorinated biphenyls in four freshwater sediments. Environ. Toxicol. Chem. 27: 2287–2295.

Swiatkowski, A., M. Pakula, S. Biniak, and M. Walczyk. 2004. Influence of the surface chemistry of modified activated carbon on its electrochemical behaviour in the presence of lead(II) ions. Carbon 42: 3057–3069.

Talley, J.W., U. Ghosh, S.G. Tucker, J.S. Furey, and R.G. Luthy. 2002. Particle-scale understanding of the bioavailability of PAHs in sediment. Environ. Sci. Technol. 36: 477–483.

Thies, J. and M. Rillig. Characteristics of biochar: biological properties. *In*: J. Lehmann, S. Joseph (eds.). 2009. Biochar for Environmental Management: Science and Technology. Earthscan.

Tordoff, G.M., A.J.M. Baker, and A.J. Willis. 2000. Current approaches to the revegetation and reclamation of metalliferous mine wastes. Chemosphere 41: 219–228.

Trakal, L., M. Komárek, J. Száková, V. Zemanová, and P. Tlustoš. 2011. Biochar application to metal-contaminated soil: Evaluating of Cd, Cu, Pb and Zn sorption behavior using single- and multi-element sorption experiment. Plant, Soil and Environment 57: 372–380.

Uchimiya, M., S. Chang, and K.T. Klasson. 2011a. Screening biochars for heavy metal retention in soil: Role of oxygen functional groups. J. Hazard. Mater. 190: 432–441.

Uchimiya, M., I.M. Lima, K.T. Klasson, and L.H. Wartelle. 2010a. Contaminant immobilization and nutrient release by biochar soil amendment: Roles of natural organic matter. Chemosphere 80: 935–940.

Uchimiya, M., I.M. Lima, K. Thomas Klasson, S. Chang, L.H. Wartelle, and J.E. Rodgers. 2010b. Immobilization of Heavy Metal Ions (CuII, CdII, NiII, and PbII) by Broiler Litter-Derived Biochars in Water and Soil. J. Agric. Food Chem. 58: 5538–5544.

Uchimiya, M., L.H. Wartelle, K.T. Klasson, C.A. Fortier, and I.M. Lima. 2011b. Influence of Pyrolysis Temperature on Biochar Property and Function as a Heavy Metal Sorbent in Soil. J. Agric. Food Chem. 59: 2501–2510.

Van Zwieten, L., S. Kimber, S. Morris, K. Chan, A. Downie, J. Rust, S. Joseph, and A. Cowie. 2010. Effects of biochar from slow pyrolysis of papermill waste on agronomic performance and soil fertility. Plant Soil 327: 235–246.

Vangronsveld, J., R. Herzig, N. Weyens, J. Boulet, K. Adriaensen, A. Ruttens, T. Thewys, A. Vassilev, E. Meers, E. Nehnevajova, D. van der Lelie, and M. Mench. 2009. Phytoremediation of contaminated soils and groundwater: lessons from the field. Environmental Science and Pollution Research 16: 765–794.

Vaughan, R.L. and B.E. Reed. 2005. Modeling As(V) removal by a iron oxide impregnated activated carbon using the surface complexation approach. Water Res. 39: 1005–1014.

Wang, H.L., K.D. Lin, Z.N. Hou, B. Richardson, and J. Gan. 2010. Sorption of the herbicide terbuthylazine in two New Zealand forest soils amended with biosolids and biochars. J. Soils Sediments 10: 283–289.

Warnock, D., J. Lehmann, T. Kuyper, and M. Rillig. 2007. Mycorrhizal responses to biochar in soil—concepts and mechanisms. Plant Soil 300: 9–20.

Warren, G.P., B.J. Alloway, N.W. Lepp, B. Singh, F.J.M. Bochereau, and C. Penny. 2003. Field trials to assess the uptake of arsenic by vegetables from contaminated soils and soil remediation with iron oxides. Sci. Total Environ. 311: 19–33.

Waychunas, G.A., C.S. Kim, and J.F. Banfield. 2005. Nanoparticulate Iron Oxide Minerals in Soils and Sediments: Unique Properties and Contaminant Scavenging Mechanisms. Journal of Nanoparticle Research 7: 409–433.

Wen, B., R.-j. Li, S. Zhang, X.-q. Shan, J. Fang, K. Xiao, and S.U. Khan. 2009. Immobilization of pentachlorophenol in soil using carbonaceous material amendments. Environ. Pollut. 157: 968–974.

Williams, P.N., A.H. Price, A. Raab, S.A. Hossain, J. Feldmann, and A.A. Meharg. 2005. Variation in Arsenic Speciation and Concentration in Paddy Rice Related to Dietary Exposure. Environ. Sci. Technol. 39: 5531–5540.

Williams, P.T. and A.R. Reed. 2004. High grade activated carbon matting derived from the chemical activation and pyrolysis of natural fibre textile waste. J. Anal. Appl. Pyrolysis 71: 971–986.

Xing, B.S. and J.J. Pignatello. 1997. Dual-mode sorption of low-polarity compounds in glassy poly(vinyl chloride) and soil organic matter. Environ. Sci. Technol. 31: 792–799.

Yamato, M., Y. Okimori, I.F. Wibowo, S. Anshori, and M. Ogawa. 2006. Effects of the application of charred bark of *Acacia mangium* on the yield of maize, cowpea and peanut, and soil chemical properties in South Sumatra, Indonesia. Soil Sci. Plant Nutr. 52: 489–495.

Yang, X.-B., G.-G. Ying, P.-A. Peng, L. Wang, J.-L. Zhao, L.-J. Zhang, P. Yuan, and H.-P. He. 2010. Influence of Biochars on Plant Uptake and Dissipation of Two Pesticides in an Agricultural Soil. J. Agric. Food Chem. 58: 7915–7921.

Yang, Y., W. Hunter, S. Tao, D. Crowley, and J. Gan. 2009. Effect of activated carbon on microbial bioavailability of phenanthrene in soils. Environ. Toxicol. Chem. 28: 2283–2288.

Yang, Y.N. and G.Y. Sheng. 2003. Enhanced pesticide sorption by soils containing particulate matter from crop residue burns. Environ. Sci. Technol. 37: 3635–3639.

Yang, Y.N., G.Y. Sheng, and M.S. Huang. 2006. Bioavailability of diuron in soil containing wheat-straw-derived char. Sci. Total Environ. 354: 170–178.

Yu, X.-Y., C.-L. Mu, C. Gu, C. Liu, and X.-J. Liu. 2011. Impact of woodchip biochar amendment on the sorption and dissipation of pesticide acetamiprid in agricultural soils. Chemosphere 85: 1284–1289.

Yu, X.Y., G.G. Ying, and R.S. Kookana. 2006. Sorption and desorption behaviors of diuron in soils amended with charcoal. J. Agric. Food Chem. 54: 8545–8550.

Yu, X.Y., G.G. Ying, and R.S. Kookana. 2009. Reduced plant uptake of pesticides with biochar additions to soil. Chemosphere 76: 665–671.

Yuan, J.H. and R.K. Xu. 2011. The amelioration effects of low temperature biochar generated from nine crop residues on an acidic Ultisol. Soil Use and Management 27: 110–115.

Zarić, Snežana D. 2003. Metal Ligand Aromatic Cation-π Interactions. European Journal of Inorganic Chemistry 2003: 2197–2209.

Zhang, H., K. Lin, H. Wang, and J. Gan. 2010. Effect of Pinus radiata derived biochars on soil sorption and desorption of phenanthrene. Environ. Pollut. 158: 2821–2825.

Zhou, Z.L., D.J. Shi, Y.P. Qiu, and G.D. Sheng. 2009. Sorptive domains of pine chars as probed by benzene and nitrobenzene. Environ. Pollut. 158: 201–206.

Zhu, D., S. Kwon, and J.J. Pignatello. 2005. Adsorption of Single-Ring Organic Compounds to Wood Charcoals Prepared under Different Thermochemical Conditions. Environ. Sci. Technol. 39: 3990–3998.

Zimmerman, A.R., B. Gao, and M.-Y. Ahn. 2011. Positive and negative carbon mineralization priming effects among a variety of biochar-amended soils. Soil Biol. Biochem. 43: 1169–1179.

Zimmerman, J.R., U. Ghosh, R.N. Millward, T.S. Bridges, and R.G. Luthy. 2004. Addition of Carbon Sorbents to Reduce PCB and PAH Bioavailability in Marine Sediments: Physicochemical Tests. Environ. Sci. Technol. 38: 5458–5464.

Zimmerman, J.R., D. Werner, U. Ghosh, R.N. Millward, T.S. Bridges, and R.G. Luthy. 2005. Effects of dose and particle size on activated carbon treatment to sequester polychlorinated biphenyls and polycyclic aromatic hydrocarbons in marine sediments. Environ. Toxicol. Chem. 24: 1594–1601.

5

Studying the Role of Biochar using Isotopic Tracing Techniques

Bruno Glaser,[1,a,]* Katja Wiedner[1,b] and
Michaela Dippold[2]

[1]Martin-Luther-University Halle-Wittenberg, Soil Biogeochemistry, von-Seckendorff-Platz 3, 06120 Halle, Germany.
[a]E-mail: bruno.glaser@landw.uni-halle.de
[b]E-mail: katja.wiedner@landw.uni-halle.de
[2]University of Bayreuth, Agroecosystem Research, 95440 Bayreuth, Germany;
E-mail: michaela.dippold@uni-bayreuth.de
*Corresponding author

Introduction

Several names are used for materials produced by pyrolysis such as pyrogenic carbon, black carbon, char(coal), biochar, pyrochar or plant char (Schmidt and Noack 2000, Glaser et al. 2002, Lehmann 2007, Bird and Ascough 2011, Knicker 2011). There are two major problems related to this variety of names. First, material derived from pyrolysis is not a defined substance, instead material properties depend on the production process, temperature and duration which is better expressed by a combustion continuum or by elemental ratios such as O/C and H/C (Fig. 1). There is still no common definition for this type of materials. Second, pyrolysis products are important across a range of disciplines such as material science, natural sciences including soil science, chemistry and biology, and agronomy each of which has an own terminology and a set of methodologies for the analysis of this material and interpretation of results. The term biochar was first used by Karaosmanoglu et al. (2000) for pyrogenic carbon produced on purpose by humans under controlled pyrolysis conditions in order to generate (i) gases for power generation (syngas), (ii) a range of bio-oils and (iii) biochar for soil amelioration in agriculture and for carbon sequestration.

Biochar has a high potential for long-term C sequestration due to its highly aromatic structure leading to a certain resistance against degradation in the terrestrial environment (Hedges et al. 2000). Also its positive effect on soil fertility especially in highly weathered soils of the humid tropics is well known from the famous *terra preta* phenomenon (Glaser et al. 2001, Glaser et al. 2002, Glaser 2007, Glaser and Birk 2012). A further complication for

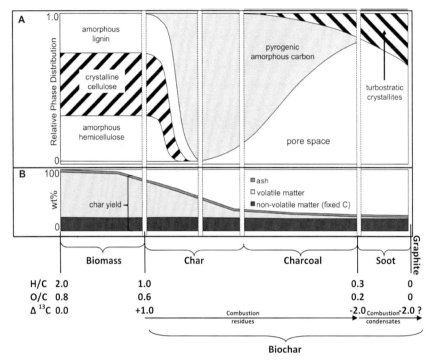

Figure 1. Combustion continuum with corresponding O/C, H/C and Δ¹³C ratios (relative to bulk parent material) and biochar window (modified from Keiluweit et al. 2010).

the study of the fate of biochar in the environment is the fact that natural (biological) processes can produce condensed aromatic structures similar to the ones of biochar (Brodowski et al. 2005, Glaser and Knorr 2008). This is the reason why sophisticated analytical tools such as isotopic techniques together with ecological pools (physical and/or chemical) are necessary to unambiguously identify the fate of biochar or other pyrolytic materials in the environment which will be discussed in this chapter.

1 Isotopes at Natural Abundance

The study of stable (e.g., ¹³C and ¹⁵N) and radioactive isotopes (e.g., ¹⁴C) in the environment have proved very useful in investigating soil carbon and nitrogen cycles (Glaser 2005) and soil trophic relationships (Staddon 2004). Recent methodological and technical advances have greatly extended the possibilities for the application of isotope techniques to terrestrial ecology and have vastly improved our knowledge of belowground ecosystem functioning and will continue to do so. These techniques are also necessary for better understanding of processes related to the fate of biochar in the environment and, thus, for a possible large-scale implementation of biochar technologies for sustainable management of natural resources and for climate change mitigation.

1.1 ^{13}C

Nuclear magnetic resonance spectroscopy (NMR) such as solid-state and liquid ^{13}C NMR is a valuable tool to characterize soil organic matter and humification processes in soil (Kögel-Knabner 1997). These techniques allow a screening of soil organic matter (SOM) quality and can differentiate between lipids (alkyl-C, 0–45 ppm), sugars (O-alkyl-C, 45–110 ppm), aromatics (lignin and condensed aromatic moieties such as biochar, 111–160 ppm) and O-containing functional groups (Carbonyl-C, carboxyl-C, 160–220 ppm) (Kögel-Knabner 2000). Several studies used the natural occurrence of ^{13}C isotope in ^{13}C NMR spectroscopy for detection and quantification of aromatic moieties derived from charred organic matter in soils. For instance, Zech et al. (1990) and Glaser et al. (2003) showed a predominance of aromatic, char-derived C in pre-Columbian Anthropogenic soils of Amazonia (*terra preta*). Downie et al. (2011) showed similar solid-state ^{13}C NMR spectroscopy proving the presence of charcoal in 650 to 1609 years old Australian Anthropogenic sites (*terra preta Australis*) and the similarity to Amazonian *terra preta*. Several studies reported higher aromatic carbon content after repeated wild fires in different ecosystems (e.g., Hammes et al. 2008, Knicker et al. 2008). Therefore, ^{13}C NMR is an appropriate tool to distinguish different C forms of SOM and show similarity of SOM quality among different ecosystems. However, a differentiation between different sources (natural vs. anthropogenic/charring vs. biological) of condensed aromatic moieties is not (yet) possible with this technique. In addition, there are problems with the quantification of aromatic carbon but special and complicated techniques (e.g., bloch decay) are available to avoid this bias (Kögel-Knabner 2000).

Basis for the work with stable isotope ratio ($\delta^{13}C$) at natural abundance is the fact that during kinetic and thermodynamic processes such as biochemical reactions, phase changes or diffusion, heavier isotopes are discriminated against the lighter counterparts because of the higher kinetic energy of the latter. As a consequence, the lighter isotopes accumulate relative to the heavier ones in the reaction products. For easier handling, the small numbers achieved when calculating the natural $^{13}C/^{12}C$ ratio and comparing the ratios of different samples, the isotope composition is expressed as $\delta^{13}C$ value, introduced by Craig (1953). Differences in $\delta^{13}C$ values of various plant species were first described by Bender (1971), which led to the discovery of three different mechanisms of photosynthetic CO_2 fixation in plants (C3, C4 and CAM). As a consequence, the range of the

δ^{13}C value in plant material varies between $-10‰$ and $-30‰$ with a mean δ^{13}C value of $-27‰$ for C3 plants and $-13‰$ for C4 plants (Glaser 2010). Therefore, the maximal amplitude for studies at natural C isotopes is 14‰. In addition, different plant compartments have a different C isotope signature due to different biochemical synthesis pathways prone to different discrimination against the heavier isotope (Hobbie and Werner 2004). Thus, the δ^{13}C value of leaf sugars is enriched by about 0–2‰ compared to bulk plant tissue. Lipids are generally depleted in ^{13}C compared other plant components by up to 16‰ (Fig. 2). High concentrations of lignin and lipids are common in woody tissue or foliage of woody plants, with little differentiation between C3 and C4 grasses in lignin and lipid concentrations. Thus, the lower lipid δ^{13}C value relative to bulk carbon is a property of many grasses and herbaceous plants.

The modification of carbon isotopes during pyrolysis is generally low with a maximum variation of up to 2‰ (Figs. 1 and 2). This low variation compared to bulk isotope signature of parent plant material can be explained by a dominance of thermodynamic processes during pyrolysis (temperature controlled) which exhibit generally low discrimination among different isotopes (Schmidt 2003). In detail, ^{13}C increases at pyrolysis temperature between 100 and 300°C were observed due to the loss of isotopically lighter lipids. Additionally, isotopically heavier cellulose-derived carbon is trapped within the initial charcoal structure. At temperature above 300°C which is

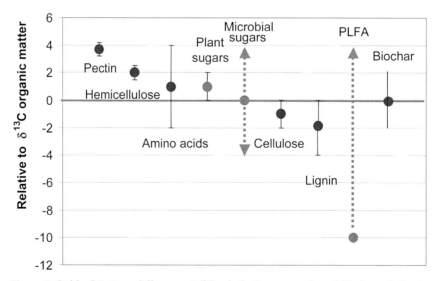

Figure 2. Stable C isotope differences (Δ^{13}C) of plant compounds and biochar relative to whole plant δ^{13}C values arbitrarily set to 0‰. Data are from Glaser (2005) and from Bird and Ascough (2011).

normal for biochar, ^{13}C decreases, probably due to the loss of isotopically heavier cellulose (Fig. 1, Krull et al. 2003, Bird and Ascough 2012). This general pattern was corroborated by a study of Czimczik et al. (2002). They charred soft- and hardwood at different temperature (150, 340 and 480°C) and examined the mass loss (7–82%), increase of the organic carbon (OC) (0–32%) and changes in ^{13}C proportion. Charring at the lowest temperature showed an enrichment of ^{13}C in all chars whereas ^{13}C was depleted at 340 and 480°C.

Along with temperature, there is evidence that $\delta^{13}C$ of char is modulated by starting material and pyrolysis conditions (Bird and Ashough 2012). Duration of charring does not induce significant $\delta^{13}C$ differences of char, but the proportion of oxygen present during pyrolysis may modulate $\delta^{13}C$ of char (Bird and Ascough 2012).

Changes in isotopic composition of biochar after deposition or biochar addition to soil is less studied. A summary of available peer-reviewed studies is given in Table 1. Changes of $\delta^{13}C$ of biochars produced from sugarcane bagasse (C4) at different temperature after three month of exposure in a C3 soil were negligible except for the biochar produced at 310°C which was exposed under elevated pH conditions (Bird and Ascough 2012). The authors suggest that the pH value plays a key role with respect to the interaction of biochar and local environment which is also confirmed by the results of Braadbaart et al. (2009).

A combination of physical fractionation procedures such as particle size or density fractionation with $\delta^{13}C$ analysis can be used to quantify turnover and stabilization of biochar in different SOM pools. Balesdent and Mariotti (1996) estimated the residence time of SOM in individual particle size fractions by the increase in $\delta^{13}C$ of each fraction over 13 years of maize cultivation. Without isotope labelling, increasing $\delta^{13}C$ values with decreasing particle size are often reported and attributed to preferential loss of ^{12}C during microbial degradation as well as the enrichment of organic compounds less available for microbial digestion (Balesdent and Mariotti 1996, Stemmer et al. 1999, Gerzabek et al. 2001). This mineralization effect might bias ^{13}C natural abundance studies when not taken into account. In the tropical environment, however, ^{13}C enrichment was only observed in the clay fraction (Roscoe et al. 2001). It is generally assumed that soil organic matter turnover time increases with decreasing particle size (Christensen 1996). However, the stability of silt and clay fractions are often controversially discussed, especially when soils contain high amounts of recalcitrant carbon such as biochar (Skjemstad et al. 1996, Glaser et al. 2001b, Glaser and Amelung 2003). Therefore, it is most likely, that biochar accumulated in the silt and clay fractions is responsible for the stability of these fractions (Glaser et al. 2001b).

Table 1. Summary of biochar studies with ^{13}C at natural abundance.

Experiment	Soil type	BC material	Duration	Results	Reference
Fate of BC on arable soil surface at 26 °C annual mean temperature, measuring CO_2 ($\delta^{13}C$)	Typic Haplustox	mango tree wood (*Mangifera indica* L.), 400-600 °C; pH value 8.74, 8.92	2 y	< 3 % biochar loss, positive and negative priming effect observed	(Major et al., 2010)
C mineralization from biochar and LOM/soil using contrasting $\delta^{13}C$ values	smectite-rich clayey soil	wood, 450, 550 °C	120 d at 20 °C	0.4-1.1 % loss of BC, 4.2-13.6 % LOM/soil; positive priming effect	(Keith et al., 2011)
CO_2 flux rates and $\delta^{13}C$ of a BC set in soils with different OM status	C3 soils; fallow, arable and grassland soils	sugarcane bagasse and sugarcane trash (C4), 350, 450, 550 °C	1 14 d at 30 °C	0.24-4.58 % loss of BC, positive and negative priming effect observed	(Cross and Sohi, 2011)
BC mineralization in five different soils measuried by CO_2 ($\delta^{13}C$) evoultion	two Alfi sols, two Entisols, Mollisol	oak (*Quercus laurifolia*), pine (*Pinus taeda*) and bubinga (*Guibourtia demeusei*), Eastern gamma grass (*Tripsacum dactyloides*), bagasse (sugar cane), 250, 400, 525, 600 °C	505 d	1.4 % loss of biochar per year, positive and negative priming effect observed	(Zimmerman et al., 2011)
Changes of $\delta^{13}C$ biochar composition in soil; elevated pH values	C3 soil	sugarcane bagasse (C4), 300, 400, 500 °C	3 month	Negligible changes; except 310 °C biochar which was under elevated pH conditions	(Bird and Ascough, 2012)

Natural abundance of stable carbon isotopes was also used to investigate the interactive priming potential of biochar and labile organic matter (LOM). Keith et al. (2011) examined the effect of LOM on biochar-C mineralization in a smectite-rich clayey soil or vice-versa. Biochar produced from *Eucalyptus salinga* at 450 and 550°C under slow pyrolysis conditions, LOM produced from sugar cane and the soil had contrasting $\delta^{13}C$ values for separating the C mineralization sources. Biochar degradation ranged only between 0.4 to 1.1% but increased with increasing LOM addition. In turn, LOM-C mineralization decreased with biochar-C addition. From these results it is obvious that biochar has a stabilizing effect on LOM in soil and is therefore suitable for long-term C-sequestration in soil.

Cross and Sohi (2011) determined the priming potential of sugarcane (C4 plant) bagasse biochars (slow pyrolysis) by incubating it in a C3 soil. They compared CO_2 flux rates and measured its corresponding $\delta^{13}C$ values (attributed to the sources) by using a simple two compartment mixing model. Results indicated that although the mineralization rate of biochar-amended soils was often higher due to the loss of labile biochar components, biochar did not prime native soil organic matter. In some cases a negative priming effect occurred in biochar-amended soils, presumably due to a stabilization effect of the labile soil organic matter at the presence of biochar (Cross and Sohi 2011).

1.2 ^{14}C

^{14}C develops constantly in the upper atmosphere through collision of neutrons of cosmic radiation and ^{14}N (Libby 1946). In contrast to the stable isotopes ^{12}C and ^{13}C, ^{14}C is radioactive and has a half-life of $5,730 \pm 40$ yr (Godwin 1962). ^{14}C oxidizes rapidly to $^{14}CO_2$ and thus, is taken up by global biospheric C reservoirs via photosynthesis (Levin 2000). The exchange stops after dying of the organism and the ^{14}C content decreases through radioactive decay. Charred organic matter may have a half-life up to 5–7 ky due to the high aromaticity and high resistance against oxidative degradation. Charcoal is therefore of high interest as a source of geochronological data by using ^{14}C isotope (Preston and Schmidt 2006, Ascough et al. 2009) despite several factors such as climate, depositions environment or land use influencing its stability (Czimczik and Masiello 2007). Using biochar produced from ^{14}C-depleted materials such as, e.g., lignite would be a means for taking advantage of ^{14}C measurements at natural abundance. However, ^{14}C dating is resource-demanding which makes this technique not appropriate for routine analysis of biochar experiments.

1.3 ^{15}N

^{15}N NMR spectra of natural organic matter are dominated by resonances from amides, peptides, indoles, lactams and carbazoles ranging from 220 to 285 ppm, with a maximum around 259 ppm. Similar spectral features were reported for compost, soils and sediments (Knicker et al. 1996). This was generally interpreted as a dominance of amide N in peptides as free amino acids would give a signal at 346 ppm. Thermal treatment leads to N-containing heterocyclic aromatic structures such as indoles, pyrroles or pyridines which give ^{15}N NMR signals between 25–145 ppm (pyridine-type compounds) and 145–240 ppm for indole- and pyrrole-type structures (Knicker 2010).

The solid-state ^{15}N-NMR spectra of biomass (*Lolium rigidum*) subjected to severe heating revealed amide-N in forms which are resistant to the thermal treatment. Progressive heating was found to occur in two well-defined stages: in the early stage the free amino acid and some NH$_2$ groups were removed, but no substantial disruption of the peptide structure was observed. In the final stage of burning the amide-N was converted to heterocyclic structures such as pyrroles, imidazoles and indoles (Knicker et al. 1996). These findings suggest that a major portion of the N is in stable forms that may survive further microbial degradation such as heterocyclic N. Nevertheless, it also appeared that most of this recalcitrant, "unknown" N forms still consists of amide structures remaining in black carbon-like soil organic matter that have been found to be relatively stable against chemical, biological and thermal degradation (Almendros et al. 1990). In conclusion, these results indicate that heating peat organic matter involves aromatization and formation of heterocyclic N. From the biogeochemical point of view, these can be considered as mechanisms leading to the transformation of labile compounds into environmentally recalcitrant forms (Alemendros et al. 2003).

Another application of natural abundance ^{15}N isotope technique (δ^{15}N) could be quantification of N retention after biochar addition as improved N recycling due to reduced N losses would lead to less ^{15}N fractionation compared to systems receiving no biochar. However, we are aware of no study in which this concept has been tested. In addition, when biochar is combined with compost as suggested recently (Fischer and Glaser 2012), δ^{15}N values should increase due to isotope fractionation during composting. Lynch et al. (2006) reported a δ^{15}N increase from 0.3 to 8.2‰ during composting and a stabilization of these δ^{15}N values during two years field incubation corroborating the assumption that compost-derived N is stabilized in soil.

1.4 ^{34}S

Four of the 25 existing sulphur isotopes are stable (^{32}S, ^{33}S, ^{34}S and ^{36}S) and could provide information about the source of sulphur (Peterson and Fry 1987). Sulphur isotopes could be useful in discriminating charred organic matter sources such as archaeological charcoals from fossil fuel-derived charring products. Bird and Ascough (2012) suggested additionally that sulphur isotopes could be useful to track interactions between biochar and the soil environment. Furthermore, sulphur can be a major source of acid rain and information about its abundance, distribution and origin is of high importance in coal combustion (Dai et al. 2002). Therefore, using the sulphur isotope analysis of aerosols is a powerful tool to identify sources of sulphur in the atmosphere, estimating the emission factors, and tracing the spread of sulphur from anthropogenic sources through ecosystems (Sinha et al. 2008). At present, no studies are available using sulphur isotopes in biochar. However, some studies performed by geologists exist about sulphur in coals investigating the spatial distribution, origin or mechanisms of sulphur incorporation into coals using an isotopic approach (e.g., Dai et al. 2002). Demir et al. (1993) determined the δ^{34}S value in the raw coals as well as in the chars and pyrolyzed the raw coals at 250–550°C to produce the chars. δ^{34}S proved a movement of the pyritic sulphur (about 18%) into the macerals after charring to 550°C. The main objective of the study was to provide more knowledge about the spatial distribution and movement of sulphur chars and coals to design more efficient processes to desulphurize or to make chemicals from coal (Demir et al. 1993). However, whether δ^{34}S analyses are also useful for biochar studies is not clear as biochar should be very low in sulphur compared to fossil coals.

1.5 ^{18}O

In disciplines like palaeoclimatology or palaeooceanography, δ^{18}O acts as a palaeoenvironmental proxy. Regarding historical charcoals and biochar research, oxygen isotopes could provide information on environment and/or conditions of formation and/or the dynamic of biochar degradation (Bird and Ascough 2012). Aging of biochar under natural conditions such as archaeological environments affects the increase of oxygen and decrease of carbon on biochar surface (Cheng et al. 2006, Cheng et al. 2008) and could be therefore suitable for providing the above-mentioned information (Fig. 3). Newly formed charcoal contains a significant proportion of oxygen and hydrogen (e.g., Ascough et al. 2010a) and O/C ratio increases over time upon environmental exposure through oxidation of the biochar surface

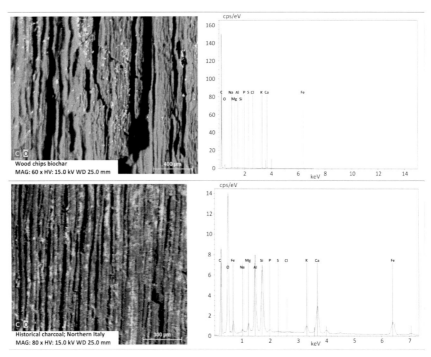

Figure 3. Scanning electron microscopy (SEM) in combination with energy-dispersive X-ray spectroscopy (EDS) images show the differences of the oxygen and carbon level between a freshly produced biochar (upper part) and a charcoal from a historical pile (lower part) seen by light grey for oxygen and black for carbon (left). EDS spectra on the right show clear enhanced surface oxidation of the historical charcoal compared with the freshly produced biochar. The O/C ratio [atom%] of the biochar sample is 0.069 compared to 1.041 of the historical charcoal (pictures: K. Wiedner 2011).

(Cheng et al. 2006, Cheng et al. 2008). Therefore, the potential exists for using $\delta^{18}O$ and δD to obtain information on environment and/or conditions of formation and/or the dynamics of biochar degradation, but there have been no studies in this area.

2 Isotope Labelling Techniques

Isotope labelling is one of the most powerful tools to investigate organic matter decomposition dynamics. With respect to biochar, it allows to specifically follow transformation and degradation processes in the soil—plant—water—atmosphere continuum. Isotope labelling of biochar allows to track the source of evolved CO_2 be it biochar or native SOM (Spokas 2010). There are several studies measuring the CO_2 production as an indicator for biochar degradation (van Zwieten et al. 2009, Kimetu and Lehmann 2010, van Zwieten et al. 2010, Augustenborg et al. 2011) but only few studies used isotopically labelled biochar (Kuzyakov et al. 2009, Steinbeiss et al. 2009, Hilscher and Knicker 2011a,b). The disadvantage in estimating biochar decay only based on unlabeled CO_2 efflux is its small contribution to CO_2 compared to SOM and it is therefore unsuitable (Kuzyakov et al. 2009). Below we will review studies that used artificial ^{13}C, ^{14}C and ^{15}N labelling techniques to investigate biochar stability and changes of microorganisms in soil ecosystems through biochar addition.

2.1 ^{13}C

Steinbeiss et al. (2009) used ^{13}C-labelled glucose (0% N) and yeast (5% N) as parent material for hydrothermal carbonization (hydrochar) to investigate its stability in an arable and forest soil and the reaction of soil microorganisms. Respiration measurements of the ^{13}C-labelled hydrochars resulted in a mean residence time of the two different hydrochars between 4 and 29 years, which is rather low compared with biochars produced by pyrolysis of an estimated half life 5–7 ky (e.g., Preston and Schmidt 2006).

Hilscher and Knicker (2011a) performed microcosm incubation experiments using ^{13}C- and ^{15}N-enriched grass-derived biochar mixed with a sub soil material taken from a Haplic Cambisol. After 1 month, up to 13.8% of the ^{13}C label and 12.4% of the ^{15}N label were detected in the mineral fraction free of particulate organic matter, indicating a fast organomineral stabilization of biochar. The chemical structure of the remaining biochar after 20 months incubation was clearly different from the initial pyrogenic material. The proportion of O-containing functional groups as revealed by ^{13}C NMR spectroscopy was increased, whereas that of aryl-C and of N-containing heterocyclic structures decreased, probably due to mineralisation and conversion to other C and N groups (Hilscher and Knicker 2011b). After 20 months of incubation, the aryl C lost up to 40% of the initial amount and up to 29% of the remaining biochar-C was assigned to carboxyl/carbonyl C and O-aryl C. These reactions alter the chemical and physical properties of the char residue and make it more available for further microbial attack but also for adsorption and cation exchange processes. This is the first study providing direct scientific evidence for the degradation of N-heterocyclic domains in biochar. Mean residence times of biochar were calculated to 3–5 years which is comparable to the results of hydrochars (Steinbeiss et al. 2009). However, due to low charring time (one and four minutes), results are not comparable with biochars normally obtained at higher temperature and longer charring time.

Watzinger et al. (2012) monitored the effects of four different ^{13}C-labelled types of biochar on soil microorganisms over one year. They performed two pot and one incubation experiment by using three different soil types (Planosol, Cambisol, Chernozem) and analyzing phospholipid fatty acids (PLFAs). The results showed that the CO_2 emission decreased within 100 days after biochar addition and only little ^{13}C was respired. They also provided evidence that microorganisms used biochar as a food source shown by the microbial PLFAs. The study showed a low degradability of the used biochar types.

A summary of available ^{13}C biochar studies is given in Table 3.

2.2 [14]C

The first scientific work with artificially ^{14}C-labelled char (biochar) degradation was conducted by Shneour (1966). He mixed the char into several sterilized and non sterilized soils (forest fire soil, forest lava soil and urban soil) and measured the ^{14}C counts per minute indicating [14]CO_2 efflux. The [14]CO_2 development was generally higher in non-sterilized soils. However, the forest lava soil and the urban soil showed a significant higher [14]CO_2 development than the forest fire soil which is a clear hint to ecosystem function and microorganism consortia (Table 4).

Bruun et al. (2008) performed a study estimating mineralization rate and assimilation of biochar by microorganisms depending on thermal alteration. Biochar was produced at different temperatures (250, 300 and 375 °C) from ^{14}C-labelled barley roots (*Hordeum vulgare*) and was incubated in a fresh sandy loam soil. Within the first 5–8 days after biochar incubation, the [14]CO_2 efflux was high but decreased thereafter. The results also showed that the mineralization rate decreased with increasing charring temperature. However, no assimilation of biochar by microorganisms could be detected. The authors concluded that either assimilation by microorganisms did not occur or the analysis methods were unsuitable (Table 4).

Hamer et al. (2004) investigated the influence of glucose addition on biochar mineralization. Furthermore, it was tested whether biochar slowed down glucose degradation due to negative priming effect. Biochar made from maize, rye residues (charring temperature 350°C) and oak wood (charring temperature 800°C) were incubated in sand for 60 days at 20°C. At the beginning, a nutrient solution an inoculum (extracted from an arable soil) containing 20 µg ^{14}C-labelled glucose per mg biochar were added. The glucose addition was repeated after 26 days. With glucose addition mineralization rate of biochar increased significantly, even more after the second addition. Due to the strong correlation between the glucose and

Table 3. Summary of biochar studies with ^{13}C labelling techniques.

Experiment	Soil type	BC material	Duration	Results	Reference
Stability of hydrochar; response of micro-organisms (PLFA)	forest and arable soil	Glucose and yeast hydrochar (HTC)	4 months	Mean residence time 4 to 29 yrs; yeast-HTC promotes fungal growth; bacteria utilized glucose-HTC	(Steinbeiss et al., 2009)
Microcosm incubation using ^{13}C- and ^{15}N labels	Haplic Cambisol	Grass biochar (BC)	20 months	Mean residence time 3-5 yrs; aryl C lost up to 40%; 29% of biochar-C was assigned to carboxyl/carbonyl C and O-aryl C	(Hilscher and Knicker 2011)
Effects of biochar on soil microorganisms; incubation an pot experiments (PLFA)	Planosol, Cambisol, Chernozem	4 different types of biochar	12 months	Low degradability of the biochar; decreased CO_2 emission after biochar addition; uptake of biochar by microorganisms	(Watzinger et al., 2012)

Table 4. Summary of biochar studies with ^{14}C labelling techniques.

Experiment	Soil type	BC material	Duration	Results	Reference
Biochar degradation in different soil types	Sterilized and non sterilized soils: forest fire soil, forest lava and urban soil	Graphite	< 60 d	$^{14}CO_2$ development higher in non-sterilized soils; forest lava and urban soil showed a significant higher $^{14}CO_2$ development	(Shneour, 1966)
BC decomposition rates based on $^{14}CO_2$ sampled 44 times during the 3.2 years	loess; Haplic Luvisol	Perennial ryegrass (*Lolium perenne*), 400 °C	3.2 y	Loss of biochar < 4.5 %; positive priming effect; half-live of biochar ~ 2000 y	(Kuzyakov et al.,1999)
Influence of glucose addition on biochar mineralization; retarding property of biochar	Sand	Maize, rye residues 350 °C; oak wood 800 °C	60 d	Increase of biochar mineralization after glucose addition; priming effect and co-metabolism	(Hamer et al., 2004)
Mineralization rate and assimilation of biochar by microorganisms	Sandy loam	Barley roots biochar (*Hordeum vulgare*); 250, 300 and 375 °C	120 d	No assimilation; Minerlization rate within the first 5-8 high; Decreasing $^{14}CO_2$ efflux with increasing charring temperature	(Bruun et al., 2008)

biochar mineralization, Hamer et al. (2004) suggested a co-metabolic process of biochar degradation triggered by microbial growth and therewith an enhanced enzyme production. It is presumed that the enhanced glucose mineralization in presence of biochar is an interactive priming effect. This leads to the assumption that if biochar promotes microbial growth in soil, degradation of labile C compounds is enhanced. The results regarding the co-metabolic process after glucose addition is confirmed by a study of Kuzyakov et al (2009). They produced biochar from ^{14}C-labelled perennial ryegrass (*Lolium perenne*). They labelled the feedstock 4 times before cutting and 3 times after cutting. They incubated the ^{14}C-labelled biochar in loess and in an Ah of a Haplic Luvisol for 3.2 years and measured ^{14}CO$_2$ evolution 44 times within this period. Additionally, they prepared two sample sets to examine the response on glucose addition and mechanical disturbance. The results showed a significant increase of biochar decomposition after adding glucose. This stimulating effect decreased after 2 weeks in soil and after 3 months in loess. The effect of mechanical disturbance was much lower in soil and loess compared to the glucose stimulation. The authors suggested a positive stimulation of microorganisms through addition of glucose and therewith a co-metabolic degradation process of biochar. The authors estimated a half-life of the examined biochar type of around 2,000 years in temperate soils. They also concluded that the use of ^{14}C-labelled biochar is a promising approach to investigate very slow processes like the microbial decomposition and chemical transformation of biochar. The C budget of biochar is calculable and biochar metabolites in SOM fractions and microbial pools identifiable (Kuzyakov et al. 2009). Nevertheless, it is still an open question if the carbon-label prefers a specific location in biochar for instance due to the higher percentage of aromatic ring structures and how far results are influenced by this or other unknown factors (Spokas 2010). A summary of available biochar ^{14}C studies is given in Table 4.

2.3 ^{15}N

Hilscher and Knicker (2011a,b) investigated the recalcitrance of pyrolzed rye grass (*Lolium perenne*) (350°C for 1 and 4 minutes) labelled during the growth phase with ^{13}C-enriched CO$_2$ gas and a ^{15}N-labeled potassium nitrate nutrient solution. They incubated the obtained biochars in subsoil material of a Haplic Cambisol for 20 months and examined biochar quality at different stages of degradation with solid-state ^{13}C and ^{15}N NMR spectroscopy. The results showed a clear increase of the proportion of O-containing functional groups whereas C and N-containing heterocyclic structures decreased. Consequences of this alteration are changes in the chemical and physical properties of biochar with increases of adsorption, cation exchange capacity and microbial access. The study contributes to the understanding of N-heterocyclic domains degradation in biochar-affected ecosystems (Table 5).

Incorporating biochar into soil (15 or 30 Mg ha^{-1}) can significantly decrease NH_3 volatilisation (up to 45%) from ^{15}N-labelled ruminant urine (Taghizadeh-Toosi et al. 2012). When the urine-treated biochar particles were transferred into fresh soil, subsequent plant growth was not affected but ^{15}N uptake into plant tissues increased, indicating that the adsorbed N was plant-available. ^{15}N recovery by roots averaged 6.8% but ranged from 26.1 to 10.9% in leaf tissue due to differing biochar properties with plant ^{15}N recovery greater when acidic biochars were used to capture ammonia. Recovery of ^{15}N as total soil nitrogen (organic+inorganic) ranged from 45 to 29% of ^{15}N applied. These results prove a synergistic mitigation option where anthropogenic ammonia emissions could be captured using biochar, and made bioavailable in soils, thus leading the capture of reactive nitrogen by crops, while simultaneously sequestering carbon in soils (Taghizadeh-Toosi et al. 2012).

Rondon et al. (2007) conducted a ^{15}N isotope dilution study with common beans in comparison to non-N-fixing beans. Biological N fixation contributed 50–72% to total N uptake and significantly improved crop performance. Biomass production and grain yield were significantly lower in non-N-fixing beans than in N-fixing beans. Nitrogen deficiency was visually apparent in non-N-fixing plants by chlorosis just 20 days after planting confirmed by chlorophyll measurements. Biochar additions significantly increased biological N fixation of common beans at all application rates. The reason for the improved biological N fixation is most likely a combination of factors related to nutrient availability in soil and stimulation of plant–microbe interactions. It is likely that reduced N availability stimulated biological N fixation as foliar N concentrations decreased with all biochar application rates, although the proportion of biological N fixation increased, the total N uptake from soil significantly decreased.

In a field trial in central Amazonia, Steiner et al. (2008) investigated the influence of biochar and compost on N retention in soil and plants. Fifteen months after organic matter admixing (0–0.1 m soil depth), ^{15}N-labelled $(NH_4)_2SO_4$ (27.5 kg N ha^{-1} at 10 atom% excess) was added to the soil. The tracer was measured in top soil (0–0.1 m) and plant samples taken at two successive sorghum (*Sorghum bicolor* L. Moench) harvests. The N recovery in biomass was significantly higher when the soil contained compost (14.7% of applied N) in comparison to only mineral-fertilized plots (5.7%) due to significantly higher crop production during the first growth period. After the second harvest, the retention in soil was significantly higher in the biochar-amended plots (15.6%) in comparison to only mineral-fertilized plots (9.7%) due to higher N retention in soil. The total N recovery in soil, crop residues, and grains was significantly ($p < 0.05$) higher on compost

Table 5. Summary of biochar studies with ^{15}N labelling techniques.

Experiment	Soil type	BC material	Duration	Results	Reference
Stability of hydrochar; response of micro-organisms (PLFA)	forest and arable soil	Glucose and yeast hydrochar (HTC)	4 months	Mean residence time 4 to 29 yrs; yeast-HTC promotes fungal growth; bacteria utilized glucose-HTC	(Steinbeiss et al., 2009)
Microcosm incubation using ^{13}C- and ^{15}N labels	Haplic Cambisol	Grass biochar (BC)	20 months	Mean residence time 3-5 yrs; aryl C lost up to 40%; 29% of biochar-C was assigned to carboxyl/carbonyl C and O-aryl C	(Hilscher and Knicker 2011)
Effects of biochar on soil microorganisms; incubation an pot experiments (PLFA)	Planosol, Cambisol, Chernozem	4 different types of biochar	12 months	Low degradability of the biochar; decreased CO_2 emission after biochar addition; uptake of biochar by microorganisms	(Watzinger et al., 2012)

(16.5%), biochar (18.1%), and biochar plus compost treatments (17.4%) in comparison to only mineral-fertilized plots (10.9%). Organic amendments increased the retention of applied fertilizer N. One process in this retention was found to be the recycling of N taken up by the crop. However, little is still known on the relevance of N immobilization, reduced N leaching, and gaseous losses as well as other potential processes for increasing N retention which should be unravelled in future studies (Steiner et al. 2008). A summary of available biochar ^{15}N studies is given in Table 5.

3 Tracing the Isotope Signature of Specific Compounds of Biochar Biogeochemistry

3.1 Condensed Aromatic Moieties (Black Carbon)

The C isotope signature of physically isolated charcoal was used to trace its origin and to reconstruct vegetation community (Krull et al. 2003, Pessenda et al. 1996a, Winkler 1994). One well-established method for biochar quantification in the environment is using benzenepolycarboxylic acids (BPCA) as molecular markers for the condensed aromatic backbone (Glaser et al. 1998). Until now, ^{13}C isotope measurement of biochar consisted of physical separation and bulk ^{13}C measurement of biochar residue after 4M trifluoroacetic acid and 65% nitric acid digestion (Glaser and Knorr 2008) or an analytical procedure involving measurement of the trimethylsilyl or methyl aromatic polycarboxylic acid derivatives by gas chromatography–combustion–isotope ratio mass spectrometry (GC-C-IRMS; Glaser and Knorr 2008). However, BPCA derivatives contain up to 150% derivative carbon, necessitating post-analysis correction for the accurate measurement of $\delta^{13}C$ values, leading to a detection limit of about 2–3‰ (Glaser and Knorr 2008). Recently, a new method using ion-exchange chromatography of underivatized BPCA by liquid chromatography–combustion–isotope ratio mass spectrometry (LC-C-IRMS) was developed (Yarnes et al. 2011).

3.2 Biologically Produced Aromatic Moieties

One complication of using ^{13}C isotope labelling in biochar research is the fact, that condensed aromatic structures present in biochar can be also produced biologically, e.g. by soil fungi (Glaser and Knorr 2008). Aspergillin, the black pigment of *Aspergillus niger* (which is ubiquitous in soils) was reported to contain condensed aromatic structures, similar to those of black carbon or biochar (Brodowski et al. 2005). Biological black carbon production in soils is a function of time and climate (Glaser and Knorr 2008). In temperate regions, annual biological black carbon production contributes between

1–3% of the soil-inherent black carbon inventory while in tropical regions this figure could reach 9% (Glaser and Knorr 2008).

3.3 Biochar Degradation Products

Biochar oxidation in the laboratory yields a suite of BPCA (Glaser et al. 1998), suggesting similar oxidation products in soils. Although the BPCA method has been applied in several studies (e.g., Glaser et al. 2000, Glaser and Amelung 2003, Dai et al. 2005, Rodionov et al. 2006, Brodowski et al. 2007, Hammes et al. 2008), little is known on the occurrence of free BPCA in soils (Haumaier 2010). They have been found to be constituents of fulvic acids from a Podzol in Canada (Khan and Schnitzer 1971, Ogner and Schnitzer 1971) and from a Podzol in Scotland (Ikeya et al. 2006), in fulvic acids from a Cambisol and an Andosol in Japan (Ikeya et al. 2006), and in alkaline extracts from some Cambisols in Thailand (Möller et al. 2000). In the latter study, the occurrence of benzene hexacarboxylic (mellitic) acid was attributed for the first time to the oxidation of biochar in soil. In order to test the assumption of the ubiquity of BPCA in soils and to evaluate their potential as tracers of biochar degradation, Haumaier (2010) developed a method for the routine determination of free BPCA in soil samples which was established and applied to a collection of samples of soils from various origins. The method involves extraction from soil samples with 0.5 M NaOH and quantification by gas chromatography–mass spectrometry.

As expected, BPCA turned out to be as ubiquitous as black carbon. They were detected not only in every soil sample investigated so far, but also in samples from drill cores up to a depth of 10 m and in recently deposited calcareous tufa. The concentration covered a range similar to that of some phenolic acids. The range exceeded those reported for low-molecular-weight aliphatic acids or simple sugars in soils. The distribution of BPCA in soil profiles indicated a considerable potential of translocation within, and export from, soil, in particular of benzene hexacarboxylic (mellitic) acid. Mellitic acid may therefore be present in almost any geochemical sample affected by seepage water from soils. Its high water solubility and strong metal-complexing ability suggest it may be involved in metal-transport processes, at least on geological timescales (Haumaier 2010). However, free BPCA concentrations cannot be used as a measure of biochar degradation because of the strong susceptibility to leaching of BPCA. In addition, low abundance in biochar-affected soils of high pH are interpreted as an indication that pH may be a regulating factor in biochar stability. The obvious mobility and polyfunctional nature in particular of mellitic acid suggest that it may be involved in many geochemical processes, the extent of its participation is still unknown. Yet, no results combining these molecular with isotopic approaches in biochar studies are known.

3.4 Polycyclic Aromatic Hydrocarbons

Polycyclic aromatic hydrocarbons (PAH) belong to an important group of organic pollutants in the environment (Wilcke 2000). Similar to black carbon, PAH are mainly produced by incomplete combustion of organic matter and fossil fuels. Additional PAH sources are diagenesis in sediments as well as microbial synthesis (Wilcke 2000). Therefore, both natural and anthropogenic PAH sources exist. Many studies deal with PAH source apportionment using C isotope signature of individual PAH in soils and sediments. Principally, the C isotope signature of pyrolysis-derived PAH resembles those of the parent material (Krull et al. 2003, Schmidt et al. 2004). Lichtfouse and Eglinton (1995) quantified the PAH contribution from fossil fuel combustion by combined compound-specific ^{13}C of individual n-alkanes and ^{14}C analysis. The $\delta^{13}C$ signature of mean n-alkanes averaged –35, –23, and –29.5‰ for C3 plants, C4 plants, and fossil fuels, respectively. Fossil fuel polluted soils were identified and pollution could be quantified by the difference between $\delta^{13}C$ values of individual n-alkanes between a contaminated and a control soil under wheat. In this way, the authors calculated that 78% of soil carbon derived from fossil fuel.

Wilcke et al. (2002) used the C isotope signature of individual PAH in temperate and tropical soils and in tropical wood and termite nests to distinguish different PAH sources. $\delta^{13}C$ values of pyrolytic PAH such as benzofluoranthenes and benzopyrenes varied between –25.3 and –24.6‰ in accordance with other literature data. While no significant difference of $\delta^{13}C$ values of naphthalin could be detected among the investigated samples, $\delta^{13}C$ values of perylene decreased in the order temperate soils (–27.0‰) > termite nests (–31.4‰) > tropical soils (–32.4‰). Although the authors concluded a significant biological perylene production from the termite nests and tropical soils results, this interpretation is ambiguous because biochar produced by forest fires which are widespread in the investigation area would produce perylene with a similar isotope signature. Further studies using ^{13}C labelling are encouraged to corroborate biological PAH production by termites.

PAH source apportionment is ambiguous when more than two possible PAH sources are involved in PAH contamination and when compound-specific stable isotope analysis is used alone. Phillips and Gregg (2001) suggested use of more complicated models, in which not only the variability of the mixed sample is considered, but also the contribution of more than two sources can be estimated. However, up to now such models have been rarely used.

In conclusion, PAH are added together with biochar to soil and could be potential biochar markers. Tracing ^{13}C isotope signature of individual PAH in soil could help for PAH source identification as indicated in Table 6.

Table 6. Natural ^{13}C isotope signature of individual polycyclic aromatic hydrocarbons (PAH) in soil samples near a Highway and a rural area with potential PAH sources (data from Glaser et al. 2005, n.d. is not determined).

TABLE 5. Carbon Isotope Signature (δ^{13}C in ‰) of Individual Polycyclic Aromatic Hydrocarbons in Soil Samples at a Major German Highway and in an Urban Area as Well as Potential Polycyclic Aromatic Hydrocarbon Sources

	distance to highway/source (m)			domestic soot	tire	coal soot[a]	wood soot[b]	car soot[b]
	1	15	500					
acenaphthene	−28.0	−27.5	−27.1	−29.7	−32.3	−25.0	−24.0	−24.2
fluorene	−25.0	−23.2	−22.2	n.a.[c]	n.a.	−25.0	−25.2	−23.5
phenanthrene/anthracene	−24.1	−20.2	−25.2	−22.9	−26.2	−24.8	−24.0	−24.2
benzo[b+j+k]fluoranthene	−33.9	n.a.	−32.9	−28.8	−29.7	−24.2	−26.5	−24.2
fluoranthene	n.a.	−23.0	−22.1	n.a.	n.a.	−25.9	−26.5	−23.5
pyrene	n.a.	n.a.	−22.1	n.a.	n.a.	−26.1	−25.2	−22.5
benzo[a]anthracene/benzo[a]pyrene	−31.3	−29.7	−29.4	n.a.	n.a.	−25.2	−24.8	−24

[a] Reference 40. [b] Reference 27. Independent from fuel type, i.e., diesel or gas. [c] n.a. not available.

3.5 Soil Microorganisms

The knowledge about the influence of biochar on soil biota is still very limited. A helpful tool for understanding the interaction between soil microorganisms and biochar is the use of Stable Isotope Probing (SIP) which is combinable with other molecular tracing techniques for examining the fluxes of carbon within the soil system (Verheijen et al. 2009). The principle of SIP is based on using substrates enriched with stable isotopes (^{13}C, ^{15}N, ^{18}O) consumed by microorganisms incorporating the heavy isotope into their biomass inclusive their DNA and rRNA (Neufeld et al. 2007). Several molecular and analytical techniques can be applied to analyze the labelled biomarkers, such as phospholipid-derived fatty acids (PLFA), ribosomal RNA and DNA for identification of microorganisms which incorporated the substrate (Neufeld et al. 2007). In addition, substrate utilization of position-specific labelled tracers can be used for metabolic tracing (Dijkstra 2011)—a highly sensitive tool to detect changes in microbial C utilization in biochar amended soils.

Ascough et al. (2010) investigated the response of two saprophytic white-rot fungi species, *Pleurotus pulmonarius* and *Coriolus versicolor*, to biochar (charcoal) as a growth substrate using a combination of optical microscopy, scanning electron microscopy, elemental abundance measurements, and isotope ratio mass spectrometry (^{13}C and ^{15}N) to investigate fungal colonisation of control and incubated samples of Scots Pine (*Pinus sylvestris*) wood, and charcoal from the same species produced at 300°C. Both fungal species colonized the surface and interior of wood and charcoals over time periods of less than 70 days; however, distinctly different growth forms are evident between the exterior and interior of the charcoal substrate, with hyphen penetration concentrated along lines of structural weakness. Although the fungi were able to degrade and metabolise the pine wood, charcoal does not form a readily available source of fungal nutrients at least for these species under the conditions used in this study (Ascough et al. 2010).

Steinbeiss et al. (2009) used ^{13}C-labeled glucose (0% N) and yeast (5% N) as parent material for hydrothermal carbonization (hydrochar) to investigate the reaction of soil microorganisms in arable and a forest soils. Changes and groups of microbial biomass were quantified and identified by the extraction of phospholipid fatty acids (PLFAs) and using the compound-specific ^{13}C isotope ratio. Results showed an enhanced fungal growth after four months promoted by yeast-derived hydrochar whereas Gram-negative bacteria utilized hydrochar made of glucose. The authors concluded that the chemical structure or the degree of condensation of hydrochars are more important regulators of microbial activity than soil properties.

3.6 Trace Gases (CO_2, N_2O, CH_4)

If biochar is rather stable in soils, only little C should be mineralized and, thus, released as CO_2 which make isotope approaches in biochar studies challenging. However, increased CO_2 production from soils after biochar amendments has been frequently reported. In addition, ^{13}C signature of CO_2 evolved during the first days of incubation was the same as the ^{13}C signature of biochar, confirming that biochar significantly contributed to CO_2 release from soil (Smith 2010). However, this effect ceased after about one week corroborating (i) that biochar consists of a labile and a stable pool comprising about 10–20% and (ii) that most of biochar decomposes only slowly (80–90%). It is likely that a fraction of the condensates from the bio-oil formed during pyrolysis absorbed to the biochar during cooling. These condensates are likely the source of the labile C pool and thus do not originate from the stable carbonized components of the biochar, since only about 10–20% of the soluble component is mineralized.

3.7 Low Molecular Weight Organic Substances

Biochar amendment is known to change the fate of different classes of low molecular weight organic substances (LMWOS) in soil. Most research focuses on degradation, leaching and sorption of agrochemical substances, e.g., pesticides in soil (Jones et al. 2011). Kookana (2012) showed that for the majority of pesticides, sorption to biochar-amended soils is increased. This not only means a reduced leaching from those soils but also a reduced efficiency. In contrast, interaction of biochar with naturally occurring organic substances is only rarely investigated. Kasoi et al. (2010) showed a sorption capacity of 1% by weight for catechol as a model substance for naturally occurring LMWOS). However, the sorption mechanisms and with it consequences for microbial utilization and stabilization of LMWOS are not yet understood. Here isotope studies based on labelling of LMWOS can enlighten mechanistic principles of this interaction: First short-term studies adding position-specific labelled alanine sorbed on active coal revealed that this sorption strongly reduced the mineralization of alanine from 50% of clay mineral-bound alanine to less than 10% of active coal-bound alanine. However, application of position-specific tracer revealed that metabolization pathway of the sorbed alanine had changed (Fig. 4).

The divergence index (Eqn. 1) shows the fate of individual C atoms from the position i within a transformation process relative to the mean transformation of the n total number of C atoms in the substance, and thus does not reflect any absolute values but just the relative ratio of the isotopomeres. A DI_i of 1 means that the transformation of this C atom corresponds to that of uniformly labelled substance (average of all C

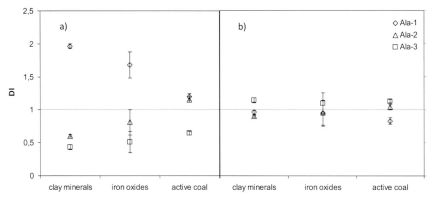

Figure 4. Position-specific divergence index DI_i (N=5, ± SE) of alanine C functional groups bound to different sorbents: a) DI of mineralized alanine C and B) DI of sorbed alanine C after 56 h.

atoms). DI_i ranges from 0 to N, and values between 0 and 1 reflect reduced incorporation of the C into the investigated pool, whereas values between 1 and N show increased incorporation of the C atom into this pool as compared to the average.

$$DI_i = \frac{n \cdot [C_i]}{\sum_{i=1}^{i=n} [C_i]} = \frac{3 \cdot [Ala - C_i]}{\sum_{1}^{3} [Ala - C_i]} \tag{1}$$

This index revealed that charcoal-bound alanine showed less discrimination between the positions compared to mineral-bound alanine. Carboxylic groups show highest stabilization by the active coal which is opposite to the behaviour of dissolved alanine and mineral-bound alanine. Whether this is a direct result of the sorption mechanism or an indirect result of a change in the microbial metabilization pathways due to lower availability of alanine has to be investigated by further experiments with LMWOS of various substance classes and biochar of different qualities. Thus, position-specific isotope labelling approaches can meet the demand to identify the underlying processes to explain the frequently observed C stabilization effects of biochar amendment.

4 Perspectives: Functional-Group Labelling

Most calculations of biochar turnover times based on labelling studies assume homogenously labelled biochar. On the one hand it is well known that biochar quality (e.g., molecular structure and isotopic composition) strongly depends on pyrolysis conditions and that preferential incorporation of distinct substance classes depending on charring temperature is one of the major reasons for this (Knicker 2007, Keiluweit et al. 2010, Bird and Ascough 2012). On the other hand, many studies showed that single pulse labelling leads to an inhomogenous distribution of the added ^{13}C, ^{14}C or ^{15}N in the different chemical components of the plants (Meharg 1994, Paterson et al. 2009). Whereas this fact might cause a bias in stability calculations of inhomogenously labelled biochar, it may, otherwise, be a perspective for producing biochar, being labelled at distinct functional groups. Such biochar would give us the prospect to better understand microbial degradation processes, initial degradation of labile biochar compounds and long-term stability of charred material in soil.

Acknowledgements

This work was partly funded by the EuroChar project of the European Community (FP7-ENV-2010 Project ID 265179) and the ClimaCarbon project of the German Ministry of Education and Research (01LY1110B).

References

Almendros, G., H. Knicker, and F.J. González-Vila. 2003. Rearrangement of carbon and nitrogen forms in peat after progressive thermal oxidation as determined by solid-state ^{13}C- and ^{15}N-NMR spectroscopy. Org. Geochem. 34: 1559–1568.

Almendros, G., F.J. González-Vila, and F. Martin. 1990. Fire-induced transformation of soil organic matter from an oak forest. An experimental approach to the effects of fire on humic substances. Soil Sci. 149: 158–168.

Ascough, P.L., C.J. Sturrock, and M.I. Bird. 2010. Investigation of growth responses in saprophytic fungi to charred biomass. Isotopes Env. Health Stud. 46: 64–77.

Ascough, P.L., M.I. Bird, F. Higham, W. Meredith, C.E. Snape, and C.H. Vane. 2009. Hydropyrolysis as a new tool for radiocarbon pre-treatment and the quantification of black carbon. Quat. Geochron. 4: 140–147.

Augustenborg, C.A., S. Hepp, C. Kammann, D. Hagan, and C. Müller. 2011. Impact of biochar on soil N_2O and CO_2 emissions in the presence of earthworms. J. Env. Qual. doi:10.2134/jeq2011.0119.

Balesdent, J. and A. Mariotti. Measurement of soil organic matter turnover using ^{13}C natural abundance. pp. 83–111. *In*: T.W. Boutton and S. Yamasaki [eds.]. 1996. Mass Spectrometry of Soils. Marcel Dekker, New York.

Bender, M.M. 1971. Variations in the $^{13}C/^{12}C$ ratios of plants in relation to the pathway of photosynthetic carbon dioxide fixation. Phytochem. 10: 1239–1244.

Bird, M.I. and P.L. Ascough. 2012. Isotopes in pyrogenic carbon: a review. Organic Isotopes in Soil Org. Geochem. 42: 1529–1539.

Braadbaart, F., I. Poole, and A.A. van Brussel. 2009. Preservation potential of charcoal in alkaline environments: an experimental approach and implications for the archaeological record. J. Archaeol. Sci. 36: 1672–1679.

Brodowski, S., W. Amelung, L. Haumaier, and W. Zech. 2007. Black carbon contribution to stable humus in German arable soils. Geoderma 139: 220–228.

Brodowski, S., W. Amelung, L. Haumaier, C. Abetz, and W. Zech. 2005. Morphological and chemical properties of black carbon in physical soil fractions as revealed by scanning electron microscopy and energy-dispersive X-ray spectroscopy. Geoderma 128: 116–129.

Bruun, S., E.S. Jensen, and L.S. Jensen. 2008. Microbial mineralization and assimilation of black carbon: Dependency on degree of thermal alteration. Org. Geochem. 39: 839–845.

Cheng, C.-H., J. Lehmann, and J.E. Thies. 2006. Oxidation of black carbon by biotic and abiotic processes. Org. Geochem. 37: 1477–1488.

Cheng, C.-H., J. Lehmann, J.E. Thies, and S.D. Burton. 2008. Stability of black carbon in soils across a climatic gradient. J. Geophys. Res. 113: 13–29.

Craig, H. 1953. The geochemistry of stable carbon isotopes. Geochim. Cosmochim. Acta 3: 53–92.

Christensen, B.T. Carbon in primary and secondary organomineral complexes. pp. 97–165. *In:* M. R. Carter and B.A. Stewart [eds.]. 1996. Advances in Soil Science, Vol. 27. CRC Lewis, Boca Raton.

Cross, A. and S.P. Sohi. 2011. The priming potential of biochar products in relation to labile carbon contents and soil organic matter status. Soil Biol. Biochem. 43: 2127–2134.

Czimczik, C.I. and C.A. Masiello. 2007. Controls on black carbon storage in soils. Global Biogeochem. Cycles 21: 1–8.

Czimczik, C.I., C.M. Preston, M.W.I. Schmidt, R.A. Werner, and E. Schulze. 2002. Effects of charring on mass, organic carbon, and stable carbon isotope composition of wood. Org. Geochem. 33: 1207–1223.

Dai, X., T.W. Boutton, B. Glaser, R.J. Ansley, and W. Zech. 2005. Black carbon in a temperate mixed-grass savanna. Soil Biol. Biochem. 37: 1879–1881.

Dai, S., D. Ren, Y. Tang, L. Shao, and S. Li. 2002. Distribution, isotopic variation and origin of sulfur in coals in the Wuda coalfield, Inner Mongolia, Chin. Intern. J. Coal Geol. 51: 237–250.

Demir, I., R.D. Harvey, and K.C. Hackley. 1993. SEM-EDX and isotope characterization of the organic sulfur in macerals and chars in Illinois Basin coals. Org. Geochem. 20: 257–266.

Dijkstra, P., J.C. Blankinship, P.C. Selmants, S.C. Hart, G.W. Koch, E. Schwartz, and B.A. Hungate. 2011. Probing carbon flux patterns through soil microbial metabolic networks using parallel position-specific tracer labelling. Soil Biol. Biochem. 43: 126–132.

Downie, A.E., L. van Zwieten, R.J. Smernik, S. Morris, and P.R. Munroe. 2011. Terra Preta Australis: reassessing the carbon storage capacity of temperate soils. Agricul. Ecosys. Env. 140: 137–147.

Fischer, D. and B. Glaser. Synergisms between compost and biochar for sustainable soil amelioration. pp. 167–198. *In:* K. Sunil and A. Bharti [eds.]. 2012. Management of Organic Waste. InTech.

Gerzabek, M.H., G. Habenhauer, and H. Kirchmann. 2001. Soil organic matter pools and carbon-13 natural abundances in particle-size fractions of a long-term agricultural field experiment receiving organic amendments. Soil Sci. Soc. Am. J. 65: 352–358.

Glaser, B. and J.J. Birk. 2012. State of the scientific knowledge on properties and genesis of Anthropogenic Dark Earths in Central Amazonia (terra preta de Índio). Geochim. Cosmochim. Acta doi:10.1016/j.gca.2010.11.029, in press.

Glaser, B. Variation of abundances of Carbon isotopes in nature. pp. 823–842. *In:* D. Beauchemin and D.E. Matthews [eds.]. 2010. Elemental and Isotope Ratio Mass Spectrometry, Vol. 5. Elsevier.

Glaser, B. and K.H. Knorr. 2008. Isotopic evidence for condensed aromatics from non-pyrogenic sources in soils—implications for current methods for quantifying soil black carbon. Rapid Commun. Mass Spectrom. 22: 935–942.

Glaser, B. 2007. Prehistorically modified soils of central Amazonia: a model for sustainable agriculture in the twenty-first century. Phil. Trans. Roy. Soc. B. 362: 187–196.

Glaser, B. 2005. Compound-specific stable-isotope (delta C-13) analysis in soil science. J. Plant Nutr. Soil Sci. 168: 633–648.

Glaser, B., A. Dreyer, M. Bock, S. Fiedler, M. Mehring, and T. Heitmann. 2005. Source apportionment of organic pollutants of a highway-traffic-influenced urban area in

Bayreuth (Germany) using biomarker and stable carbon isotope signatures. Env. Sci. Tech. 39: 3911–3917.

Glaser, B. and W. Amelung. 2003. Pyrogenic carbon in native grassland soils along a climosequence in North America. Glob. Biogeochem. Cyc. 17: 1064, doi: 10.1029/2002 GB002019.

Glaser, B., G. Guggenberger, W. Zech. Organic chemistry studies on Amazonian Dark Earths. pp. 227–241. *In:* J. Lehmann, D. Kern, B. Glaser, and W. Woods [eds.]. 2003. Amazonian Dark Earths: Origin, Properties, and Management. Kluwer, Dordrecht, The Netherlands.

Glaser, B., J. Lehmann, and W. Zech. 2002. Ameliorating physical and chemical properties of highly weathered soils in the tropics with charcoal—a review. Biol. Fertil. Soils 35: 219–230.

Glaser, B., L. Haumaier, G. Guggenberger, and W. Zech. 2001a. The 'Terra Preta' phenomenon: a model for sustainable agriculture in the humid tropics. Naturwiss. 88: 37–41.

Glaser, B., G. Guggenberger, and W. Zech. Black carbon in sustainable soils of the Brazilian Amazon region. pp. 359–364. *In:* R.S. Swift and K.M. Spark [eds.]. 2001b. Understanding & Managing Organic Matter in Soils, Sediments & Waters. International Humic Substances Society, St. Paul, MN.

Glaser, B., E. Balashov, L. Haumaier, G. Guggenberger, and W. Zech. 2000. Black carbon in density fractions of anthropogenic soils of the Brazilian Amazon region. Org. Geochem. 31: 669–678.

Glaser, B., L. Haumaier, G. Guggenberger, and W. Zech. 1998. Black carbon in soils: the use of benzenecarboxylic acids as specific markers. Org. Geochem. 29: 811–819.

Godwin, H. 1962. Half-life of radiocarbon. Nature 195: 984.

Hamer, U., B. Marschner, S. Brodowski, and W. Amelung. 2004. Interactive priming of black carbon and glucose mineralisation. Org. Geochem. 35: 823–830.

Hammes, K., R.J. Smernik, J.O. Skjemstad, and M.W.I. Schmidt. 2008. Characterisation and evaluation of reference materials for black carbon analysis using elemental composition, colour, BET surface area and ^{13}C NMR spectroscopy. Appl. Geochem. 23: 2113–2122.

Haumaier, L. 2010. Benzene polycarboxylic acids—A ubiquitous class of compounds in soils. J. Plant Nutr. Soil Sci. 17: 727–736.

Hedges, J.I., G. Eglinton, P.G. Hatcher, D.L. Kirchman, C. Arnosti, S. Derenne, R.P. Evershed, I. Kögel-Knabner, J.W. de Leeuw, R. Littke, W. Michaelis, and J. Rullkötter. 2000. The molecularly uncharacterized component of nonliving organic matter in natural environments. Org. Geochem. 31: 945–958.

Hilscher, A. and H. Knicker. 2011a. Carbon and nitrogen degradation on molecular scale of grass-derived pyrogenic organic material during 28 months of incubation in soil. Soil Biol. Biochem. 43: 261–270.

Hilscher, A. and H. Knicker. 2011b. Degradation of grass-derived pyrogenic organic material, transport of the residues within a soil column and distribution in soil organic matter fractions during a 28 month microcosm experiment. Org. Geochem. 42: 42–54.

Hobbie, E.A. and R.A. Werner. 2004. Intramolecular, compound-specific, and bulk carbon isotope patterns in C-3 and C-4 plants: a review and synthesis. New Phytol. 161: 371–385.

Ikeya, K., Y. Ishida, H. Ohtani, and A. Watanabe. 2006. Effect of off-line methylation using carbanion and methyl iodide on pyrolysis-gas chromatographic analysis of humic and fulvic acids. J. Anal. Appl. Pyrolysis 75: 174–180.

Jones, D.L., G. Edwards-Jones, and D.V. Murphy. 2011. Biochar-mediated alterations in herbicide breakdown and leaching in soil. Soil Biol. Biochem. 43: 804–813.

Karaosmanoglu, F., A. Isigigur-Ergundenler, and A. Sever. 2000. Biochar from the straw-stalk of rapeseed plant. Energy Fuels 14: 336–339.

Kasozi, G.N., A.R. Zimmerman, P. Nkedi-Kizza, and B. Gao. 2010. Catechol and Humic Acid Sorption onto a range of laboratory-produced black carbons (biochars). Env. Sci. Techn. 44: 6189–6195.

Keiluweit, M., P.S. Nico, M.G. Johnson, and M. Kleber. 2010. Dynamic molecular structure of plant biomass-derived black carbon (biochar). Env. Sci. Techn. 44: 1247–1253.

Keith, A. and B. Singh. 2011. Interactive priming of biochar and labile organic matter mineralization in a smectite-rich soil. Env. Sci. Techn. 5: 9611–9618.

Khan, S.U. and M. Schnitzer. 1971. Further investigations on the chemistry of fulvic acid, a soil humic fraction. Can. J. Chem. 49: 2302–2309.

Kimetu, J.M. and J. Lehmann. 2010. Stability and stabilisation of biochar and green manure in soil with different organic carbon contents. Aust. J. Soil Res. 48: 577.

Knicker, H. 2011. Pyrogenic organic matter in soil: Its origin and occurrence, its chemistry and survival in soil environments. Quat. Int. doi:10.1016/j.quaint.2011.02.037: in press.

Knicker, H. 2010. "Black nitrogen"—an important fraction in determining the recalcitrance of charcoal. Org. Geochem. 41: 947–950.

Knicker, H., A. Hilscher, F.J. Gonzalez-Vila, and G. Almendros. 2008. A new conceptual model for the structural properties of char produced during vegetation fires. Org. Geochem. 39: 935–939.

Knicker, H. 2007. How does fire affect the nature and stability of soil organic nitrogen and carbon? A review. Biogeochem. 85: 91–118.

Knicker, H., G. Almendros, F.J. González-Vila, F. Martin, and H.D. Lüdemann. 1996. ^{13}C- and ^{15}N-NMR spectroscopic examination of the transformation of organic nitrogen in plant biomass during thermal treatment. Soil Biol. Biochem. 28: 1053–1060.

Kookana, R.S. 2012. The role of biochar in modifying the environmental fate, bioavailability, and efficacy of pesticides in soils: a review. Aust. J. Soil Res. 48: 627–637.

Kögel-Knabner, I. 2000. Analytical approaches for characterizing soil organic matter. Org. Geochem. 31: 609–625.

Kögel-Knabner, I. 1997. ^{13}C and ^{15}N NMR spectroscopy as a tool in soil organic matter studies. NMR in Soil Science. Geoderma 80: 243–270.

Krull, E.S., J.O. Skjemstad, D. Graetz, K. Grice, W. Dunning, G. Cook, and J.F. Parr. 2003. ^{13}C-depleted charcoal from C4 grasses and the role of occluded carbon in phytoliths. Org. Geochem. 34: 1337–1352.

Kuzyakov, Y., I. Subbotina, H. Chen, I. Bogomolova, and X. Xu. 2009. Black carbon decomposition and incorporation into soil microbial biomass estimated by ^{14}C labelling. Soil Biol. Biochem. 41: 210–219.

Lehmann, J. 2007. Bio-energy in the black. Front. Ecol. Env. 5: 381–387.

Levin, I. and V. Hesshaimer. 2000. Radiocarbon a unique tracer of global carbon cycle dynamics. Radiocarbon 42: 69–80.

Lichtfouse, E. and T.I. Eglinton. 1995. ^{13}C and ^{14}C evidence of pollution of a soil by fossil fuel and reconstruction of the composition of the pollutant. Org. Geochem. 23: 969–973.

Libby, W.F. 1946. Atmospheric helium three and radiocarbon from cosmic radiation. Phys. Rev. 69: 671–672.

Lynch, D.H., R.P. Voroney, and P.R. Warman. 2006. Use of ^{13}C and ^{15}N natural abundance techniques to characterize carbon and nitrogen dynamics in composting and in compost-amended soils. Soil Biol. Biochem. 38: 103–114.

Major, J., J. Lehmann, M. Rondon, and C. Goodale. 2010. Fate of soil-applied black carbon: downward migration, leaching and soil respiration. Global Change Biol. 16: 1366–1379.

Meharg, A.A. 1994. A critical review of labelling techniques used to quantify rhizosphere carbon flow. Plant Soil 166: 55–62.

Möller, A., K. Kaiser, W. Amelung, C. Niamskul, S. Udomsri, M. Puthawong, L. Haumaier, and W. Zech. 2000. Relationsips between C and P forms in tropical soils (Thailand) as assessed by liquid-state ^{13}C- and ^{31}P-NMR spectroscopy. Aust. J. Soil Res. 38: 1017–1035.

Neufeld, J.D., M.G. Dumont, J. Vohra, and J.C. Murrell. 2007. Methodological considerations for the use of stable isotope probing in microbial ecology. Microb. Ecol. 53: 435–442.

Ogner, G. and M. Schnitzer. 1971. Chemistry of fulvic acid, a soil humic fraction, and its relation to lignin. Can. J. Chem. 49: 1053–1063.

Paterson, E., A.J. Midwood, and P. Millard. 2009. Through the eye of the needle: a review of isotope approaches to quantify microbial processes mediating soil carbon balance. New Phytol. 184: 19–33.

Pessenda, L.C.R., R. Aravena, A.J. Melfi, E.C.C. Telles, R. Boulet, E.P.E. Valencia, and M. Tomazello. 1996a. The use of carbon isotopes (^{13}C, ^{14}C) in soil to evaluate vegetation changes during the Holocene in central Brazil. Radiocarbon 38: 191–201.

Peterson, B.J. and B. Fry. 1987. Stable isotopes in ecosystem studies. Ann. Rev. Ecol. Systemat. 18: 293–320.

Phillips, D.L. and J.W. Gregg. 2001. Uncertainty in source partitioning using stable isotopes. Oecologia 127: 171–179.

Preston, C.M. and M.W.I. Schmidt. 2006. Black (pyrogenic) carbon: a synthesis of current knowledge and uncertainties with special consideration of boreal regions. Biogeosciences 3: 397–420.

Rodionov, A., W. Amelung, L. Haumaier, I. Urusevskaja, and W. Zech. 2006. Black carbon in the Zonal steppe soils of Russia. J. Plant Nutr. Soil Sci 169: 363–369.

Rondon, M.A., J. Lehmann, J. Ramírez, Juan, and M. Hurtado. 2007. Biological nitrogen fixation by common beans (*Phaseolus vulgaris* L.) increases with biochar additions. Biol. Fertil. Soils 43: 699–708.

Roscoe, R., P. Buurman, E.J. Velthorst, and C.A. Vasconcellos. 2001. Soil organic matter dynamics in density and particle size fractions as revealed by the ^{13}C/^{12}C isotopic ratio in a Cerrado's oxisol. Geoderma 104: 185–202.

Schmidt, T.C., L. Zwank, M. Elsner, M. Berg, R.U. Meckenstock, and S.B. Haderlein. 2004. Compound-specific stable isotope analysis of organic contaminants in natural environments: a critical review of the state of the art, prospects, and future challenges. Anal. Bioanal. Chem. 378: 283–300.

Schmidt, H.L. 2003. Fundamentals and systematics of the non-statistical distributions of isotopes in natural compounds. Naturwiss. 90: 537–552.

Schmidt, M.W.I. and A.G. Noack. 2000. Black carbon in soils and sediments: Analysis, distribution, implications, and current challenges. Global Biogeochem. Cycles 14: 777–793.

Shneour, E.A. 1966. Oxidation of graphitic carbon in certain soils. Science 151: 991–992.

Sinha, B.W., P. Hoppe, J. Huth, S. Foley, and M.O. Andreae. 2008. Sulphur isotope analyses of individual aerosol particles in the urban aerosol at a central European site (Mainz, Germany). Atmos. Chem. Phys. 8: 7217–7238.

Skjemstad, J.O., P. Clarke, J.A. Taylor, J.M. Oades, and S.G. McClure. 1996. The chemistry and nature of protected carbon in soil. Aust. J. Soil Res. 34: 251–271.

Smith, J.L., H.P. Collins, and V.L. Bailey. 2010. The effect of young biochar on soil respiration. Soil Biol. Biochem. 42: 2345–2347.

Spokas, K.A. 2010. Review of the stability of biochar in soils: predictability of O:C molar ratios. Carbon Management 1: 289–303.

Staddon, P.L. 2004. Carbon isotopes in functional soil ecology. Trends in Ecology and Evolution 19:148–154.

Steinbeiss, S., G. Gleixner, and M. Antonietti. 2009. Effect of biochar amendment on soil carbon balance and soil microbial activity. Soil Biol. Biochem. 41: 1301–1310.

Steiner, C., B. Glaser, W.G. Teixeira, J. Lehmann, W.E.H. Blum, and W. Zech. 2008. Nitrogen retention and plant uptake on a highly weathered central Amazonian Ferralsol amended with compost and charcoal. J. Plant Nutr. Soil Sci. 171: 893–899.

Stemmer, M., M. von Lützow, E. Kandeler, F. Pichlmayer, and M.H. Gerzabek. 1999. The effect of maize straw placement on mineralization of C and N in soil particle size fractions. Eur. J. Soil Sci. 50: 73–86.

Taghizadeh-Toosi, A., T.J. Clough , R. R. Sherlock, and L.M. Condron. 2012. Biochar adsorbed ammonia is bioavailable. Plant Soil 350: 57–69.

van Zwieten, L., S. Kimber, S. Morris, A. Downie, E. Berger, J. Rust, and C. Scheer. 2010. Influence of biochars on flux of N_2O and CO_2 from Ferrosol. Soil Res. 48: 555–568.

van Zwieten, L., B. Singh, S. Joseph, S. Kimber, A. Cowie, and K.Y. Chan. Biochar and emissions of non-CO_2 greenhouse gases from soil. pp. 227–249. *In:* J. Lehmann and S. Joseph [eds.]. 2009. Biochar for Environmental Management—Science and Technology. Earthscan, London.

Verheijen, F.G.A., S. Jefferey, A.C. Bastos, M. van der Velde, and I. Diafas. 2009. Biochar application to soils–A critical scientific review of effects on soil properties, processes and functions. EUR 24099 EN. Office for the Official Publications of the European Communities, Luxembourg.

Watzinger, A.S. Feichtmair, F. Rempt, E. Anders, B. Wimmer, B. Kitzler, S. Zechmeister-Boltenstern, M. Horacek, F. Zehetner, S. Kloss, S. Richoz, and G. Soja. 2012. The effect of biochar amendment on the soil microbial community—PLFA analyses and 13C labeling results. Geophysical Research Abstracts 14, EGU2012-1584-4, 2012, EGU General Assembly 2012.

Wilcke, W., M. Krauss, and W. Amelung. 2002. Carbon isotope signature of polycyclic aromatic hydrocarbons (PAHs): Evidence for different sources in tropical and temperate environments? Env. Sci. Techn. 36: 3530–3535.

Wilcke, W. 2000. Polycyclic aromatic hydrocarbons (PAHs) in soil—a Review. J. Plant Nutr. Soil Sci. 163: 229–243.

Winkler, M.G. 1994. Sensing plant community and climate change by charcoal-carbon isotope analysis. Ecoscience 1: 340–345.

Yarnes, C., F. Santos, N. Singh, S. Abiven, M.W.I. Schmidt, and J.A. Bird. 2011. Stable isotopic analysis of pyrogenic organic matter in soils by liquid chromatography-isotope-ratio mass spectrometry of benzene polycarboxylic acids. Rapid Commun. Mass Spectrom. 25: 3723–3731.

Zech, W., L. Haumaier, and R. Hempfling. Ecological aspects of soil organic matter in tropical land use. pp. 187–202. *In:* P. McCarthy, C.E. Clapp, R.L. Malcolm, and P.R. Bloom [eds.]. 1990. Humic Substances in Soil and Crop Sciences. Selected Readings. American Society of Agronomy and Soil Science Society of America, Madison Wisconsin, USA.

Zimmerman, A.R., B. Gao, and M.-Y. Ahn. 2011. Positive and negative carbon mineralization priming effects among a variety of biochar-amended soils. Soil Biol. Biochem. 43: 1169–1179.

6

Designing Specific Biochars to Address Soil Constraints: A Developing Industry

Stephen Joseph,[1,*] Lukas Van Zwieten,[2,a]
Chee Chia,[3,b] S. Kimber,[2] Paul Munroe,[3,c] Yun Lin,[3,d]
Chris Marjo,[4,e] James Hook,[4,f] Torsten Thomas,[5,g]
Shaun Nielsen,[5,h] S. Donne[6] and Paul Taylor[7]

Authors' affiliations given at the end of the chapter.

Introduction

Application of biochar produced in open fires or simple kilns mixed with mineral and organic matter is an ancient practice (Lehmann and Joseph 2009). In Japan and China the smoke produced in simple kilns is condensed and refined (known as wood viniger), and then used with the biochar or as a foliar spray. Considerable data on impact of plant growth has been generated from studying the traditional practices in the Amazon, Asia and Africa (Joseph et al. 2010). Data have been published on the chemical composition and physical structure of what are known as black carbon (BC) particles that come from soils where biochar has been applied in the past (Liang et al. 2006).

Over the past 5 years there has been a large volume of data published on the application of biochar to a range of different soils. Many of the earlier pot and field trials were carried out using biochar of either unknown origin or produced in simple kilns where control of both time and temperature were either nonexistent or very basic. Application rates varied from 1 to 50 tonnes/ha. Plant responses have varied although the greatest increase in yields has been reported at application rates greater than 2 tonnes/hectare (Chan et al. 2007). More recently research work has been carried out on biochars that have been produced in reactors where process conditions were controlled and these biochars have been fully characterized. Examination of the changes in the structural and chemical composition of biochar surfaces applied to soils has indicated that biochars become more effective after the surfaces have been oxidized and organo-mineral phases are formed on the surface (Cheng 2006, Joseph et al. 2010). This research has provided guidelines for increasing the beneficial effects of biochar in the first year of application.

This chapter will summarise the available data on the properties of biochars made from diverse feedstocks at various temperatures, and give a brief overview of how the different biochars can affect yields from different crops in different soils. The chapter will summarise the methods of traditional production of biochar, the material characteristics of different biochars, and how they may change when applied to soils. Using the understanding gained from the above a number of techniques to enhance the efficacy of biochars before application to soils will be explored, including:

1) Pretreating the biomass before pyrolysis with acids, alkalis, minerals and/or clays, including different types of biomass with high and low ash content.
2) Treating after pyrolysis with acids, alkalis, minerals, other organic waste, and germination/growth promoting substances.
3) Biological treatment in the presence or absence of oxygen.
4) Addition of chemical fertilizers into the pores of the biochar.

1 The Physical and Chemical Nature of Biochars

Biochar consists of a complex mixture of organic carbon based compounds and mineral phases (Fig. 1). Metals and non-metals can coexist within the carbon matrix (Amonette and Joseph 2009). The mineral matter content of wood derived biochars is usually very small (<5wt.%) whereas those which derive from crop residues and manures is high (>10%).

1.1 Complexity of the Carbon Matrix

The internal structure of most biochars produced under 550°C consists of an irregular amorphous carbon lattice with regions of high and low condensation. Keiluweit et al. (2010) have proposed that a complex transformation of the organic structures occurs as the pyrolysis temperature increases from 200 to 700°C. From X-ray diffraction and spectroscopic analysis they suggest the existence of four distinct categories of biochars

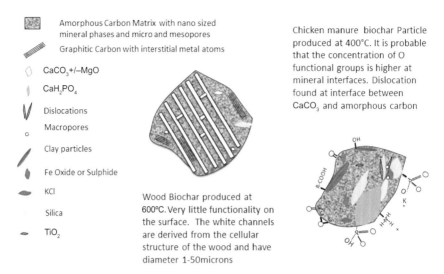

Amorphous Carbon Matrix with nano sized mineral phases and micro and mesopores

Graphitic Carbon with interstitial metal atoms

CaCO₃+/–MgO

CaH₂PO₄

Dislocations

Macropores

Clay particles

Fe Oxide or Sulphide

KCl

Silica

TiO₂

Chicken manure biochar Particle produced at 400°C. It is probable that the concentration of O functional groups is higher at mineral interfaces. Dislocation found at interface between CaCO₃ and amorphous carbon

Wood Biochar produced at 600°C. Very little functionality on the surface. The white channels are derived from the cellular structure of the wood and have diameter 1-50microns

Figure 1. Schematic of complex structure of low temperature biochars.

Color image of this figure appears in the color plate section at the end of the book.

consisting of unique mixtures of chemical phases and physical states: (i) in transition chars, the crystalline character of the precursor materials is preserved; (ii) in amorphous chars, the heat-altered molecules and incipient aromatic polycondensates are randomly mixed; (iii) composite chars consist of poorly ordered graphene stacks embedded in amorphous phases; and (iv) turbostratic chars are dominated by disordered graphitic crystallites.

A range of functional groups exist on the surface of biochars. Lower temperature biochars have a much higher concentration of oxygenated functional groups. Low temperature biochars produced under 450°C have a relatively high concentration of water and organic soluble volatile compounds. Using Liquid Chromatography with Organic Carbon Detection (LC-OCD) Lin et al. (2012b) identified that the main constituents of the water soluble organic compounds from woody biochars produced at low temperatures had characteristics that defined them as low molecular weight neutral molecules along with building blocks (the breakdown products of humic substances) with minor amounts of low molecular weight acids and biopolymers (Huber et al. 2011). Calvelo Pereira et al. (2011) identified higher molecular weight organic compounds consisting of a mixture of phenols, guaiacols and catechols, as being derived from lignin. These compounds could equate to building block compounds identified using LC-OCD. Other compounds that have been identified but not accurately quantified include alkanes, alcohols, aldehydes, acids, ketones, triols, diols, phenols, levoglucosan (Calvelo Pereira et al. 2011, Graber et al. 2011, Chia et al. 2012 (submitted)). Higher temperature biochars have much smaller concentrations of organic molecules but are dominated by low molecular weight acids (especially carboxylic) as well as low molecular weight neutrals. Small quantities of benzene, toluene, napthalene and various other polyaromatic hydrocarbons have also been identified (Calvelo Pereira et al. 2011).

During the pyrolysis process, unpaired electrons (radicals) can form in high concentration comparable to a standard DPPH molecule (2,2-diphenyl-1-picrylhydrazyl) that has a high radical content (Bourke 2007). High concentrations of radicals can lead to the chemisorption of oxygen resulting in spontaneous combustion of biochars at temperatures below 100°C. High concentration of radicals can also induce complex reactions between soil organic matter and minerals at the surface of the biochar (Bartlett 1999). Defects (e.g., dislocations and vacancies) in the carbon lattice have been reported by Bourke (2007) and can also form in areas between graphene sheets and amorphous carbon, and between the various types of carbon structures and the mineral matter. Metal and non-metal atoms can be intercalated within the carbon lattice (Ammonette and Joseph 2009). Thus biochars produced under 500°C are semiconductors and most biochars produce above 600°C are conductors (Ishihara 1996, Antal and Grønli 2003). Little data are available on electrical and electronic properties of biochars produced between 500 and 600°C (Joseph et al. 2010).

1.2 Mineral Phases

Mineral matter (amorphous and crystalline) can exist as separate micro and nano phases (oxides, sulphates/sulphides, carbonates, chlorides and phosphates) within the organic lattice. Some of these mineral phases are conductors, some semiconductors, or insulators. Fe and Fe/Mn phases can be magnetic or paramagnetic. The boundaries between the mineral phase and the organic phases have a complex structure. Joseph et al. (2010) noted that there is a high concentration of dislocations between the amorphous carbon phase and the calcium carbonate phase in chicken manure biochar produced at 350°C. There were also nanopores between these phases. Chia et al. (2012b) have noted that there can be a high concentration of functional groups containing oxygen in the carbon lattice at the interfacial boundary between the mineral phase and the organic phase.

1.3 Properties of Biochars as a Function of Process Conditions and Feedstock

Research work has illustrated that biochars produced from different feedstock and under different process conditions have different properties. The following is a summary of the main findings.

1. Characteristics of low temperature biochars (<400°C)
   a) High concentration of oxygenated functional groups that include carboxylic acids, ketones, aldehydes, hydroxyl, quinones, lactone, carboxyl anhydride, cyclic peroxide, 2H-pyran, furan, 2H-pyran-2-one (-pyrone), 2(5H)-furanone (Amonette and Joseph 2009, Keiluweit et al. 2010).
   b) High radical content and can chemisorb oxygen (Amonette and Joseph 2009). Radicals form as depolymerisation of the organic molecules in the biomass occurs.
   c) Relatively high concentration of water and organic soluble compounds, that probably improves germination, stimulates microbial growth and induces systemic resistance and hormesis (Elad et al. 2010, Graber et al. 2010). Compounds that have been identified include alkanes, alcohols, aldehydes, acids, ketones, triols, diols, phenols, levoglucosan.
   d) Low pH for wood biochars (<7.5) and relatively lower pH for sludge and manure biochars than those made at higher temperatures (Singh and Singh 2010, Rajkovich et al. 2011).
   e) High adsorption of ammonia (especially bamboo biochar) due to the high surface negative charge (Meinjie 2004) but low adsorption of some other gases (McLaughlin et al. 2009).
   f) Higher availability of nutrients for high mineral ash materials (Shinogi 2004).
   g) Relatively low water holding capacity (180mg/gm) (Krull et al. 2010).
   h) Little loss of mineral matter due to volatilisation (Enders et al. 2012).
   i) Electrical Conductivity (EC) is variable and depends on feedstock. It may be higher or lower than biochars made from the feedstock at higher temperatures (Shinogi 2004).
   j) Variable adsorption of heavy metals (depends on type of biochar and soil). Uchima et al. (2011) observed that low temperature biochars had greatest adsorption of heavy metals in a ferrosol.
   k) Decrease in methane emissions from certain biochars in certain soils (Feng et al. 2011).

2. Characteristics of low to medium temperature biochars (400–500°C)
   a) Crop residues and grasses have a high Cation Exchange Capacity (CEC) and water holding capacity, which could be attributed to the clay attached to the feedstock (Krull 2010).
   b) Significant loss of N and K from manure biochars as these are very volatile, however this is not the case for woody biochars and for some other crop residues (Rajkovich 2011, Enders et al. 2012).

c) Considerable concentration of organic compounds and acid and basic functional groups on the surfaces of wood biochars below 450°C.

d) Considerable increase in surface area and pore volume above 450°C especially for woody based biochars (Downie et al. 2009).

e) Singh and Singh (2010) noted that low temperature wood biochar (400°C) had lower CEC than high temperature wood biochar (550°C). However acid washing the biochars resulted in the decrease in CEC for both the high and low temperature biochars (Fig. 2). Krull et al. (2010) noted that the CEC of grass and crop residues at 400–450°C was significantly greater than woody biochars.

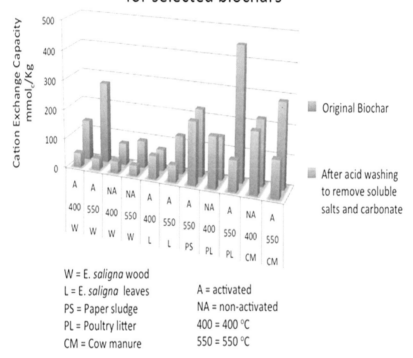

Figure 2. Plot of CEC versus feedstock, temperature and processing technique (Plot of data presented by Singh and Singh (2010).

Color image of this figure appears in the color plate section at the end of the book.

f) Phase and compositional changes in this temperature region mean that rates and amount of adsorption and desorption of nutrients, herbicides and pesticides can be variable and are a function of the type of biochar as well as soil type being used (Kookana et al. 2011).

g) Adsorption of some heavy metals appear to be lower in this temperature range (Uchima et al. 2011) although this could be a function of the soil into which the biochar is placed.

3. Characteristics of high temperature biochars (>500°C)
   For woody biochars
   a) High adsorptivity of dense organic molecules (McLaughlin 2009).
   b) High adsorptivity of heavy metals by some biochars (Namgay et al. 2010).
   c) High surface area and volume of pores below 100 nm diameter (Downie et al. 2009).
   d) High volume of pores above 100 nm diameter (Downie et al. 2009).
   e) High stability and low H/Corg ratio (Enders et al. 2012).
   f) High pH (usually >8.5) for all types of biochar (Singh and Singh 2010).
   g) Small amount of volatiles, but among them relatively high percentages of Benzene, Toluene, Ethylbenzene, and Xylene (BTEX), and possibly Polycyclic Aromatic Hydrocarbons (PAH) (Calvelo Pereira et al. 2011).
   h) For high mineral ash biochars, melting of phases can occur, especially above 550°C (Amonette and Joseph 2009). In these conditions nitrogen can be incorporated into the lattice at around 600°C as pyridine, changing the properties of the biochar (Schnitzer et al. 2007).

1.4 Effect of Biochar Properties on Crop Yields: A Brief Summary

Recent research carried out by Rajkovich et al. (2011) indicates a very complex relationship between crop yields of corn, and biomass feed stocks (8 different woods, crop residues, manure and sludges) used to make the biochar, final temperature of pyrolysis (330 vs. 500°C) and biochar application rate (0.2, 0.5, 2.0 and 5.0%). Feedstock caused the greatest variation in growth with greatest response from chicken manure biochar. The authors concluded that biochar application rates above 2.0% generally did not improve corn growth. There was a decrease in growth when biochars produced from dairy manure, paper sludge or food waste were applied at high application rates. At a biochar application rate of 0.2%, crop N uptake was 15% greater than the fully fertilized control, however there was a decrease in N uptake at higher application rates. There was not a correlation between volatile matter or ash content and crop yields or N uptake. pH had only a weak positive relationship with growth at intermediate application rates. Rajkovich et al. (2011) also reported that greater nutrient contents (N, P, K, Mg) improved growth at low application rates of 0.2 and 0.5%, but Na reduced growth at high application rates of 2.0 and 7.0% in the studied fertile Alfisol.

Verheijen et al. (2010) did a meta-analysis of the published data and found increases in yields in acidic and neutral pH soils, and in soils with a coarse or medium texture. They suggested that yield increase might be a liming effect and an improved water holding capacity of the soil, along with improved crop nutrient availability. Poultry litter biochar gave the greatest response of all of the feedstocks and biosolids a negative response.

Graber et al. (2010) found that adding a low temperature biochar made from citrus wood at different levels (0, 1, 3, 5 weight %) in optimally fertigated (fertilized and irrigated) greenhouse pot experiments resulted in an increase in leaf area and canopy dry weight in tomato and pepper plants grown in biochar amended soils, and moreover, a significant increase in pepper bud, flower and fruit yields. The rhizosphere of the biochar-amended pepper plants had significantly greater abundances of culturable microbes belonging to prominent soil-associated groups.

1.5 Implications for Developing More Plant Effective Biochars

The review above highlights the affect of feedstock and process conditions on the specific physical and chemical properties of biochars properties. The different properties can affect plant yield and resistance to disease. As well, a biochar produced from a certain feedstock and at a certain temperature may result in different plant responses in different soils. Careful blending of different biochars produced under various conditions may result in a synergistic affect that leads to greater plant yields and long-term benefits to soil health. Blends would be based on an understanding of plant nutrient and water requirements, environmental conditions as well as the chemical, physical and microbial properties of the soil.

2 An Overview of Indigenous and Historical Practices for Producing and using Biochar Mixes

2.1 Australian Aborigines

Possibly the earliest anthropogenic production of biochar for increasing availability of protein was undertaken by the Australian Aborigines (Bird et al. 2008). Fire stick farming involved lighting smoky fires when grass and scrub were moist to produce widespread concentrations of biochar (Krull et al. 2010). This type of farming also directly increased the food supply for the aborigines, by promoting the growth of bush potatoes and other edible ground level plants (Jones 1969). Aborigines also made oven mounds that comprised charcoal, minerals, organic matter and clay. Downie et al. (2010) noted that the soil in mounds had significantly greater total N, P, K and Ca content than the surrounding soils. This was also reflected in the higher mean CEC of 31.2 cmol $(^+)$ kg^{-1} and higher pH by 1.3 units, compared to the adjacent soils.

2.2 Amazon in Brazil

Amazonian Dark Earths, locally known in Brazil as Terra Preta de Indio, are biochar-rich, highly fertile soils. They occur in patches that generally range in size from several hundred to several thousand square meters, and up to 2 m deep, in a landscape of acidic and poor soils of the Amazon basin. They lie where Indigenous villages existed up to 3000 years ago. Glaser and Birk (2012) concluded from a review that "it is evident that Terra Preta is the product of inorganic (e.g., ash, bones–esp. fish) and organic (e.g., biomass wastes, manure, excrements, urine, and biochar) amendments to infertile Ferralsols. These ingredients were microbially metabolized and stabilized by humification in soil, fungi playing a bigger role in this process compared to bacteria in surrounding ecosystems. Biochar is a key component for this process due to its stability and its enrichment in terra preta. It is still unclear if terra preta was produced intentionally or unintentionally".

Major et al. (2005) found that maize yield in weeded plots was as much as 63 times greater on Dark Earth (0.55 t/ha) than on corresponding adjacent soil (9 kg/ha), and single location averages varied from 0 to 3.15 t/ha for Dark Earth.

Steiner (2007) has provided a detail description of the various techniques used to enhance soil fertility. One technique involves collecting, mounding and drying woody biomass. It is then burnt resulting in soils having a strong smell of tars. The soils are left for 3 weeks before planting and remain fertile for 3 years after this treatment. Only undemanding crops like manioc, pineapple, and trees are planted directly on these soils. For vegetables and medicinal plants further soil management is carried out. All available types of OM (organic matter) at the settlement are used (bones, wood, leafs, chicken manure), as well as certain species of rotten tree trunks. This is then reduced in an open fire before it is mixed with the soils that have already been burnt.

A range of techniques have been used to examine the particles extracted from Amazonian Dark Earth (ADE). Observations by Glaser et al. (2000), Lehmann et al. (2005), Brodowski et al. (2006) and Liang et al. (2006) revealed that the dark particles consisted of an inner phase rich in black carbon (BC) that was intimately surrounded by organo-mineral phases. Liang et al. (2006) noted that the high carbon phase consisted of highly aromatic or only slightly oxidized organic carbon. Novotny et al. (2007) found that different organic matter extracts from ADE have a higher content of aryl and ionisable oxygenated functional groups compared to adjacent soils.

Solomon et al. (2007) concluded that the organic carbon material in particles could be the products of:

i) primary recalcitrant biomolecules from organic matter in the soil or
ii) "secondary processes involving microbial mediated oxidative or extracellular neoformation reactions of soil organic carbon from BC and non-BC sources; and stabilized through physical inaccessibility to decomposition agents due to sorption onto the surface or into porous structures of the BC particles, selective preservation or through intermolecular interactions involving clay, mineral nanoparticles, and BC particles" (Solomon et al. 2007).

Detailed microscopic analysis of these particles has been carried out by Chia et al. (2011). They were found to be heterogeneous, with different mineral phases in the one particle, including calcium carbonate and phosphate, clay, iron, manganese and titanium oxides. The high amorphous carbon phase was surrounded by a layer of clay, and oxides of iron, titanium, manganese oxide nano-sized particles. All of these particles were held together by an organic phase that had significant calcium content (possibly calcium polysaccharides) (Fig. 3). The interface between the high carbon phase and the organomineral phase had a high concentration of macro and mesopores. Mossbauer spectroscopy revealed that both Fe^{2+} and Fe^{3+} existed in the matrix and magnetic susceptibility measurements showed that the particles exhibit super-paramagnetic behavior (Joseph et al. 2010). Cyclic voltametry and electrochemical impedance spectroscopy indicate that these particles are redox-active, which could result in faster decomposition of organic matter and nutrient take up by plants (Joseph et al. 2010).

Based on this information Chia et al. (2011) concluded that redox reactions between the negative surface of the biochar, organic matter and redox active minerals, especially iron and manganese could result in the formation of these organo-mineral phases. It is also possible that the formation of some of the pores at the BC/mineral interfaces is due to the formation of CO_2.

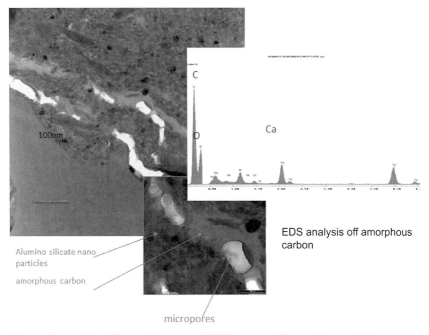

EDS analysis off amorphous carbon

Alumino silicate nano particles

amorphous carbon

micropores

Figure 3. Transmission Electron Microscopy (TEM) image and Energy Dispersive Spectroscopy (EDS) spectrum of interface between amorphous carbon and organomineral phase in a Terra Preta particle.

Color image of this figure appears in the color plate section at the end of the book.

2.3 Africa

There have been large deposits of Dark Earths found in Africa. *Ankara*, the practice of burning vegetable material that has been buried in soil beds or mounds is common among the food growers of the Western and North-Western Highlands of Cameroon (Lyonga 1980).

Urs Guyer[1] describes the *Ankara* method:

"In the NW-Province of Cameroon, when the elephant grass is completely dry, the people start to clear the plots to prepare the farms for the planting season. The farm plots are arranged in ridges, so the dry grass is placed in between the old ridges. Then the grass is covered with a rather thin layer of soil taken from the old neighboring ridges. At certain distances a hole is left open to aerate the grass layer below the soil. Then the grass is lit from the two ends of the ridge. Under the soil the grass is transformed to char. Depending on the quantity of grass the fire can burn

[1]Personal communication, Urs Guyer, Switzerland

for several days. When the soil is cold it is mixed with the char. Planting starts when the rains begin to fall. This can be weeks or even months after the *Ankara* was made".

Osinamea and Meppeb (1999) examined the soil after the formation of biochar in the Ankara soils and concluded there was a reduction in the clay-sized fraction by approximately 50% while the silt-sized fraction increased by 30%. Soil P, K, and N increased, which resulted in large increases in maize yields (6 times larger than where residues were buried without burning, and 3 times larger than surface burning).

2.4 Asia

Villagers in Northern Vietnam have developed a technique of soaking wood and bamboo in the clay and mineral rich sludge at the bottom of ponds at the edge of rice paddies (Fig. 4a). A significant volume of biochar and ash remains when this artificially aged biomass is burnt in an open fire (Fig. 4b). This layer of minerals has the effect of reducing the volume of volatiles that are liberated at once and promotes combustion around the wood. The biochar and ash is then used in home gardens to grow vegetables (Fig. 4c). Examination of the biochar using a scanning electron microscope indicated that the minerals have attached to the surface and bonded in the pores of the biochar. Ash and biochar from the burning of straw on fields is also used in conjunction with NPK fertilizer. This practice is over 40 years old and women claim that it increases the plant yield and reduces requirements for NPK fertilizer. A core sample was taken from one of the gardens and biochar was detected to a depth of 100 mm (Joseph et al. 2012).

2.5 Key Features of Traditional Practices that can Assist in Developing More Plant Effective Biochars

The above analysis of traditional practices highlight the following:

1. Biochars are often made with a mixture of different feedstocks. In a number of cases this includes human and food waste.
2. Soil rich in organic matter, clay and minerals is often adhering to the biomass feedstocks or is used to cover pyrolysing biomass.
3. Pyrolysis of biomass in these piles occurs over a range of temperatures (350–600°C) where small amounts of organic compounds (smoke) can recondense on the biochar and clay and mineral surfaces that are in the low temperature areas. The biochars that are produced have variable concentration of surface labile organic molecules that could play a significant role in improving both plant germination, and nutrient uptake (Hayashi 1990, Ishigaki 1990, Ishii et al. 1990, Dixon 1998, Light et al. 2009, Graber et al. 2010).

Figure 4. a. Aging bamboo and wood in a pond; b. Burning aged wood and bamboo in open fire; c. Biochar from aged wood produced in open fire on field.

Color image of this figure appears in the color plate section at the end of the book.

4. Mineral matter from the soil appears to become incorporated in the biochar during the pyrolysis process. Clay in the soil can give off volatiles (including HCl and HF) that assist in the pyrolysis of biomass (Heller-Kallai 1997). This can lead to the enhancement of the biochar through increased functional groups on the carbon surfaces. Chia et al. (2012b) found that when clay and iron are deposited on biochar and heated there is an increase in cyclic anhydride and carboxylic groups.
5. Biochar is applied in small amounts to the soil every year before planting.
6. In some areas the biochar is allowed to age for at least 3 weeks before planting. Ogawa (2010) states that this practice ensures that microorganisms can break down any toxic substances that are produced through condensation of the volatiles in the smoke.

3 Biochars Aged in Soils and their Relevance to Design of Specific Biochars

3.1 Understanding the Aging Process

Designing biochars that can be applied at rates that produce positive returns to farmers requires an understanding of what changes occur after their addition to soil. Interactions between biochar, soil, microbes and plant roots may occur within a short period of time after application to the soil (Joseph et al. 2010). The extent, rates and implications of these interactions, however, are far from being understood.

Biochar can oxidise if stored in a moist environment (Joseph et al. 2010). If biochar is part of a compost mix then a range of biotic and abiotic reactions can take place depending on whether the process is aerobic or anaerobic (Yoshizawa et al. 2007, Dias et al. 2010).

Different methods of incorporation may have different affects on rate of decomposition of the biochar (Kuzyakov et al. 2009). Ploughing results in greater soil mechanical disturbance than other methods such as surface application of granules, deep banding or direct drilling (Joseph et al. 2010). Mechanical disturbance of soils amended with biochar has been shown to promote biochar decomposition possibly through increasing both microbial and chemical oxidation. As Joseph et al. (2010) note this could be partly responsible for the evolution of biochar-derived CO_2 from soils within the first two weeks after amendment, which thereafter tends to decrease exponentially with time (Singh and Cowie 2008, Kuzyakov et al. 2008, Hilscher et al. 2009).

The presence of water, especially with a high content of minerals and microorganisms will have a major role in transformation of the biochar within the first month of application. Biochars start to degrade through processes such as dissolution, hydrolysis, carbonation and decarbonation, hydration and redox reactions, as well as interactions with soil biota (Shinogi et al. 2004, Major et al. 2009, Joseph et al. 2010). pH in the soil around the biochar can increase if there is dissolution of salts (e.g., K and Na carbonates and oxides). Precipitation of mineral salts may also occur when the soil dries out and the ionic activity in solution increases. These reactions probably occur preferentially within small biochar pores (Downie et al. 2009), as they have slow water percolation.

Lin et al. (2012) noted that a layer of an Al compound formed on the surface of both a chicken-manure and paper-sludge biochar after one year in the soils. A high Fe content was also found on the surface of the chicken-manure biochar that could be associated with a redox reaction between the Fe in the soil and the labile component of the biochar. For the paper-sludge biochar, they concluded that adsorption of organic matter from the soil facilitated interactions between the biochar and mineral phases in the soil. Calcium is believed to be important in this process.

Since biochar has a significant quantity of labile organic molecules, many oxidation reactions take place and these happen relatively quickly once the biochar has been added to the ground (Contescu et al. 1998). A range of oxygenated functional groups can be formed which can then react with redox active elements (Fe and Mn) to form complexes on the surface of the biochar (Nguyen and Lehmann 2009).

Biochar surfaces can initially have acidic as well as basic properties (Cheng et al. 2008, Amonette and Joseph 2009), which have an important influence both on the wetting ability of biochar particles and on the CEC (Boehm 2001). The concentration of basic sites decreases through oxidative processes as the biochar particle weathers (Cheng et al. 2008, Cheng and Lehmann 2009). Pignatello et al. (2006) have measured increases in adsorption of organic molecules including residual herbicides and pesticides (Smernik 2009, Spokas and Reicosky 2009) after the addition of biochar to soils. As organic and mineral matter builds up on the surface of the biochar, the type of compounds involved in surface reactions, the diversity and concentration of different microorganisms and the adsorption rates of soil organic matter will change (Kwon and Pignatello 2005) (see also the "Biochar-fungi interactions" chapter).

Figure 5 indicates root hairs that have penetrated into the pore of an aggregate, which is composed of a biochar particle, clay and soil minerals, as shown in the associated EDS elemental mapping. A few Fe-rich and Ti-rich particles were noted. Calcium was found present along the biochar surface, possibly representing the presence of calcium phosphate. Of particular interest is the osmium staining, which may indicate that microbes exist on the surface of the biochar particle.

Based on the literature related to organic matter, biochar and mineral interactions it can be hypothesized that (Joseph et al. 2010):

i) Surface hydrophobic and hydrophilic interactions of biochar, organic compounds and clay minerals, can occur following the conceptual model of organo-mineral interactions proposed by Kleber et al. (2007);

ii) Electron donor acceptor interactions can occur between aromatic compounds (including biochar) and mineral surfaces, as well as between two aromatic compounds (Keiluweit and Kleber 2009);

iii) Soluble organic compounds released from the biochar particles or from other organic matter in soil can become intercalated within 2:1 and 1:1 clay minerals (Lagaly 1984, Matusik et al. 2009);

iv) Complexes can form between deprotonated multidentate organic acids and transition metals (Violante and Gianfreda 2000) and also silicic acid (Marley et al. 1989).

Due to the high concentration of macropores with a diameter greater than one micron a high water-potential gradient can be generated between the inner and outer part of the pore channels (Hillel 2004) after a rain event. Capillary forces can draw soil solution containing small mineral and organic

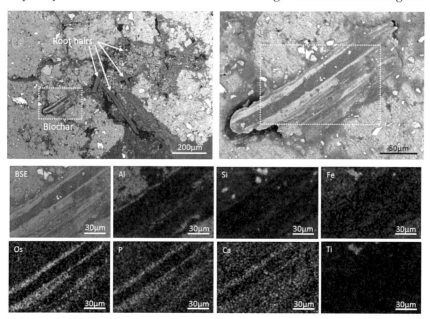

Figure 5. Aged biochar with root hair and microorganisms inside pore. Source: Electron Microscope Unit, University of NSW.

Color image of this figure appears in the color plate section at the end of the book.

particles into the pore channels. Reactions in pores can be very intricate given the heterogeneous nature of the surfaces of the solid biochar, diverse mineral and organic phases/compounds, and complexity of the dissolved organic and inorganic matter in the liquid phase that surround the solids (Hammes and Schmidt 2009).

Biochar changes its surface structure and chemical composition through interactions with a range of microorganisms. These microbes might dissolve mineral matter through exuding acids and enzymes (Gilbert and Banfield 2005) or they might decompose organic matter adsorbed on the biochar surface and within pores (Zackrisson et al. 1996).

Liang et al. (2010) observed that the total concentration of microorganisms, their diversity, and the relative proportions of fungi, bacteria, archaea and protozoa can change after addition of biochar. Within the biochar itself, facultative aerobes that inhabit the inner macropores of a biochar particle may begin to use anaerobic respiratory pathways. However, biochar addition to soil may increase soil aeration because of its highly porous nature, and hence decrease the total amount of anaerobic microsites per unit volume of soil.

Joseph et al. (2010) have shown that once the root system encounters a biochar particle, root hairs can enter the water-filled macropores or bind to the biochar surface (Fig. 5). Biochars and minerals associated with biochar particles can adsorb organic compounds released from the growing roots, including low-molecular weight organic compounds (free exudates), high-molecular-weight gelatinous material (mucilage), and sloughed-off cells and tissues and their lysates (Violante and Gianfreda 2000). These compounds can bind either to the mineral layers on the biochar or directly to the biochar surfaces. Joseph et al. note that biochar might change the complex redox potential existing in this zone, altering reactions involving free radicals. These redox reactions include transformations of C, N and S, which can involve redox reactions with Fe and Mn species.

3.2 Implications for the Design of Biochars

The rate at which a specific biochar degrades to provide both mineral and organic compounds that can be utilized for plant growth and resistance to disease is a function of the properties of the biochars as well as method of incorporation, the physical and chemical properties of the soil, the composition of the local microbial population, the number of rain events and the number of biochar particles that are in contact with roots and root hairs.

For woody biochars oxidation of surfaces appears to be the main reactions that take place after incorporation. For low temperature woody biochars labile organic molecules are dissolved or are utilized for food by microorganisms (Ogawa 2010). This is then followed by formation of complex organo-mineral phases within the pores and on the surfaces. During this time the microbial communities can change their population structure.

High mineral-ash manure biochars appear to be able to release significant amounts of nutrients in the rhizosphere after a rain event and when in contact with root hairs. For low temperature high mineral ash biochars, labile organic molecules are dissolved or are utilized as food by microorganisms. After the initial release of nutrients oxidation of the surfaces appear to take place and then a complex organo-mineral phase can be formed that may result in an increase in the local cation exchange capacity (Joseph e al. 2010).

These findings indicate that an enhanced biochar can be produced by adding nutrients to woody biochars after suitable physical, microbial or chemical surface activation. Alternately the biomass can be activated, minerals added and then pyrolysed.

4 Enhancing Properties of Biochars and Producing Blends to Meet Soil and Plant Requirements

The above examinations of traditional practices and the reactions of biochars once in soil provide guidance to methods efficient at enhancing their beneficial properties. The following is a summary of both the published literature and the experimental work that is being undertaken.

4.1 Blending Biochars

Very little research has been carried out in this area. Soils are heterogeneous; plants, and microorganisms that assist in their growth, utilize a range of organic and inorganic compounds. Thus it is probable that different soils, and different crops planted in a particular soil, will require specific combinations of biochars to optimize yields. Combinations of biochars for specific applications that may provide the range of properties needed for particular soils and plants include:

1) For soils with a low pH and a high degree of compaction for a crop with a high P, K and N requirement—a mixture of a high temperature woody biochar with a particle equivalent diameter of 3 to 8 mm, high temperature straw biomass for potassium, and relative high CEC and a low temperature chicken manure biochar for N and P.

2) For soils with a high pH, a high degree of compaction, requirement for small amounts of a nitrogen fertilizer—a mixture of low temperature wood biochar, straw biochar and a sludge containing a high nitrogen content may be appropriate.

4.2 Alkaline Pre and Post Treatment

Hina et al. (2010) have demonstrated that surface charge and density of functional groups in biochars from eucalyptus and pine bark increased after treating the feedstocks with alkaline tannery waste. Biochars were produced at 550°C from pine bark waste (PI) or from eucalyptus (EU). They were pre-treated with diluted or undiluted alkaline tannery waste. Biochars prepared from diluted and undiluted EU showed greater changes in their chemical characteristics than those made from PI. Hina et al. (2010) reported that "the specific surface area of all the biochars decreased with alkaline treatments, although acidic surface groups increased. In subsequent sorption experiments, treated biochars retained greater NH_4^+ from a 40 mg N/L waste stream (e.g., 61% retention in control EU and 83% in undiluted EU). Desorption was low, especially in treated biochars relative to untreated biochars (0.1–2% vs. 14–27%)".

Another method used to increase the concentration of surface functional groups and K content involves soaking biomass in KOH and then heating to around 700–800°C. Alternately biochars produced at 400–500°C can be post treated by reacting with KOH at ambient conditions and heating to between 220°C and 400°C (Ahmadpour and Do 1996, Illn-Gomez et al. 1996, Lozano-Castello et al. 2001).

Recent work (Lin et al. 2012) on treatment of biochars with KOH has shown changes in the:

i)   Concentration of surface acid groups,
ii) Increase in the type and concentration of water soluble labile organic molecules, and
iii) An intercalation of K into the carbon matrix and increase in the formation of K_2CO_3.

As yet agronomic trials have not been reported on the effects of KOH pretreatment of biochars on plant yields.

4.3 Acid Pre and Post Treatment

Doydora et al. (2011) found that the addition of HCl acidified biochars from pine chips and peanut hulls reduced NH_3 losses by 58 to 63% with surface-applied poultry litter and by 56 to 60% with incorporated poultry litter.

Phosphoric acid (PA) has been used to both activate the surfaces of the biochar and to add phosphorus that is plant available. Experimental work in Costa Rica (Benjamin et al. 2011) found that the properties of a biochar manufactured from Melina wood changed considerably when activated with a 10% solution of PA (Table 1). As expected the pH decreased and the P content increased. The EC and K decreased and Ca increased. Complex reactions appear to be taking place between the acid and the mineral phases. They found that this activated biochar improved the yields of maize compared with biochar that were not treated (Table 2 and 3). They also found a greater increase in yields after coating the acid treated biochar with clay and minerals and baking at 220°C to produce a biochar mineral complex (BMC) (see next section).

Table 1. Some agronomic and chemical properties of phosphoric acid treated biochar.

Source	Percentage (%)						mg/Kg of biochar				
	N	P	Ca	Mg	K	S	Fe	Cu	Zn	Mn	B
Melina Biochar	0.35	0.03	1.10	0.13	0.74	0.02	2596	4	38	31	16
Melina Biochar with P Acid	0.36	1.50	5.33	0.15	0.32	0.05	3248	11	51	187	28

Source	Moisture Content	pH	EC
	%	H_2O	mS/cm
Melina Biochar	7	9.10	0.61
Melina Biochar with P Acid	11	7.61	0.34

Table 2. Dry weight (g) for corn plants in San Juan ($p<=0.05$, there is no significant difference between any of the top 3 listed treatments, but there is relative to the bottom 3).

Treatment	Mean	N
10-30-10	143.1	15
BMC+Urea	126.3	15
Biochar+P+Urea	120.1	15
Biochar+Urea	59.7	15
Control	55.6	15
Urea	41.4	15

Table 3. Dry weight (g) for corncob in San Juan ($p<=0.05$, there is no significant difference between the top 3 treatments, but there is relative to the bottom 4).

Treatment	Mean	N
10-30-10	131.3	15
BMC+Urea	109.1	15
Biochar+P+Urea	87.4	15
Urea	50.3	15
Biochar+Urea	49.1	15
Control	45.8	15

4.4 Pretreatment of Biomass with Minerals

In section 1 we have already discussed the traditional techniques that have been used to pretreat biomass before either pyrolysis or combustion. In 2007 research work was undertaken by the University of New South Wales to attempt to mirror the processes developed by the Amazonian Indians and others. A mixture of kaolinite and bentonite (50/50), crushed brick, basaltic rock dust, lime, bones, calcium phosphate, chicken manure, and sawdust was heated to 220 and 240°C. At 220°C, the sawdust did not pyrolyse but was torrefied (depolymerisation of the biomass with subsequent loss of chemical water, CO_2 and a small amount of volatile organic compounds). The yield loss was approximately 22%, due to loss of water in both clay and the biomass, and to loss of low molecular weight organic compounds from the biomass. At 240°C the rate of thermal decomposition increased and weight loss was approximately 36% indicating the onset of pyrolysis.

Both products were analysed using a range of microscopic and spectroscopic techniques. Structures similar to those observed in *terra preta* soils were found (Chia et al. 2010).

Independent field trials were carried out by the Western Australian Department of Agriculture using this enhanced biochar (EB) (Blackwell et al. 2009). The trials were conducted in a poor, sandy soil, where wheat plants were grown in tubes inserted into field soil. PVC tubes, 10 cm in diameter, were driven into the soil vertically, with the base of the tube open to the natural subsoil. The amounts of biochar and EB available were insufficient for extensive field plots and it was believed that the in-ground tubes technique would yield results that were more representative of large-scale field trials than pot trials in a greenhouse. The experiment also included a treatment of jarrah wood biochar produced at 600°C in a retort used for making metallurgical-grade charcoal. All materials were incorporated at a rate of 10 t/ha (air dry weight) with or without adding appropriate synthetic fertilizers (n=4). At planting, a base phosphorus (P) fertilizer (single superphosphate) was applied at 100 kg/ha to all tubes. Fertilizer treatments consisted of ammonium nitrate (28 kg/ha of nitrogen and 26 kg/ha of potassium as potash).

The results of these field trials (Blackwell et al. 2009) showed that the EB applied alone gave yields similar to those obtained by applying biochar plus synthetic fertilizer, and the greatest yields were obtained when the EB was applied along with synthetic fertilizer (Fig. 6).

Chen et al. (2011) prepared three novel magnetic biochars (MOP250, MOP400, MOP700) by chemical co-precipitation of Fe^{3+}/Fe^{2+} on orange peel powder and subsequently pyrolysing under different temperatures (250°C, 400°C and 700°C). Analysis of the biochar/Fe composites by Fourier Transformed Infra-Red spectroscopy (FTIR) and separately

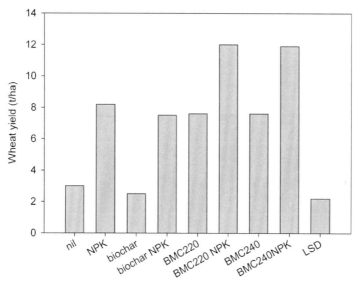

Figure 6. Results of enhanced biochars field trials: wheat yields in Western Australia. Notes: Overall high yields are an artifact of the methodology, where plants grew inside tubes. Nil = no organic amendment. NPK = synthetic fertilizer. BMC220 = BMC heated at 220°C temperature. LSD = least significant difference. Differences between treatment bars that are smaller than this bar are not statistically different, and therefore no difference can be inferred between the two treatments.

Transmission Electron Microscopy (TEM) revealed the formation of Fe_2O_3 magnetic nanoparticles on the surface of the biochar. They found that at 400°C, the composite adsorbed more organic molecules (napthalene and p-nitrotoluene) than the composites manufactured at the other temperatures and than standard biochar. The composite produced at 250°C also adsorbed more P.

4.5 Post Treatment with Organic Matter, Minerals or Chemical Fertilisers

A range of organizations are now starting to market biochar complexes. A limited amount of detailed information is available on the composition and the results of field trials for these different formulations.

One company (Biotechnology Co, Anhui) mixes NPK with biochar and a liquid clay binder in a Vapour Electro-heat Roller Dryer to produce a granule. Total Nutrient content is varied from 40–45% with $N:P_2O_5:K_2O$ ratios of 15:15:15 and 18:11:11 depending on crop and soil.

Figure 7. Biochar + NPK processing and Energy plant Sanli Co Ltd situated near Shangqiu City.

The Bamboo Research Institute has developed an organic granulated fertilizer that has high N and smaller P and K amounts, made from fermented biomass, amino acids, extra minerals, and biochar. The fertilizer is sold for approximately $A 300–350/t.

Sanli Group has developed an integrated plant that purchases local agricultural residues and converts these into biochar, wood vinegar and electricity. The biochar is mixed with chemical fertilizer elements (NPK), which is then formed into granules.

Ogawa and Yambe (1986) mixed bark charcoal with 1% (w/w) of inorganic fertilizer (N-P-K (8-8-8)) comprising urea, super lime phosphate, ammonium sulphate, and rapeseed meal and applied this fertilizer at application rates of 500 g/m^2 and 1500 g/m^2 onto fields where soya bean were grown. Plots with inorganic chemical fertilizer (applied rates of 100 g/m^2 and 200 g/m^2) were prepared. The soybean yield from the plots with charcoal fertilizers applied at 500 g/m^2 were similar to those from the control plots of only chemical fertilizer.

The biochar manufacturing company Anthroterra Pty Ltd has developed a patented process that involves activating the surface of the biochar with an oxidizing agent (phosphoric acid) and then baking of minerals and manures or sludge on the surface of biochar at temperatures between 150°C and 240°C. The composition of these complexes can be altered to meet both crop nutrient requirements (e.g., increase P and N for sweet corn) and soil requirements (e.g., changing pH and EC). Peer reviewed articles concerning the effects of this enhanced biochar on crop nutrition have not been published yet but presentations at conferences (Blackwell et al. 2009, Joseph and Solaiman 2011) indicate that improvements in yields can be achieved at 100–200kg/ha.

Figure 8 summarises the results of independent pot trials carried out by the University of Western Australia. Sandy loam soil (pH 4.8 in $CaCl_2$ at an 1:5 ratio, organic carbon 10.3 g/kg, NO_3-N 7.0 mg/kg, NH_4-N 5.0 mg/kg and Colwell P 20.0 mg/kg) was mixed with the enhanced biochar at 0, 100 or 200 kg/ha. Basal superphosphate was applied to the top soil (100mm deep) at 15 kg/ha. The soil and soil-EB mixtures (2 kg) were potted and six pre-soaked seeds of wheat (*Triticum aestivum* L. var. 'Wyalkatchem') were sown in each pot and thinned to three plants (the pot size used can support the growth of 3 plants) one week after germination. The treatments were replicated 3 times. Concerning the different biochars, B5 and B6 were manufactured from a low temperature *Acacia saligna* biochar (380°C) and B7 and B8 from a high temperature Jarrah biochar (600°C). Yield increases compared with the control (P fertilization) of approximately 40% were measured. These were attributed to an increase in mycorrhizal root colonization and overall microbial population increase (data not shown). The New South Wales Department of Primary Industry has carried out larger field trials with this enhanced biochar over a period of 3 years trialing a sweet corn summer crop and a barley winter crop. Preliminary data have been presented at a number of conferences (Joseph et al. 2011b). The results to date indicate that these enhanced fertilizers can produce similar crop responses to those produced by superphosphate and urea at application rates of less than 500kg/ha.

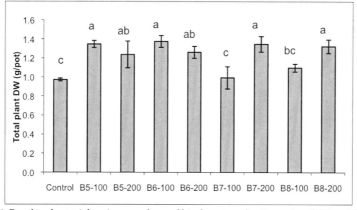

Figure 8. Results of pot trials using an enhanced biochar at application rates of 100 and 200 kg/ha using a low temperature biochar (B5 and B6) and high temperature biochar (B7 and B8).

4.6 Aerobic and Anaerobic Composting with Biochar

Extensive research has been carried out on the affect of addtion of biochar into fresh organic matter that is then composted (Fischer and Glaser 2012). They note that addition of biochar to compost improves oxygen availability and hence stimulates microbial growth and respiration rates together with moisture retention (Steiner et al. 2011). Biochar itself undergoes oxidation, improving its ability to retain nutrients, and making them plant available. Jindo et al. (2012) found that the addition of small quantities of biochar (2%) increased the degree of humification and caused significant changes in the chemical properties of the bulk organic matter (increases in alkyl-C/O-alkyl and alkyl-C/N-alkyl ratios and the aromatic index). Hua et al. (2009) also found that adding sludge biochar reduced volatilization of N.

Fischer and Glaser (2012) reported that in a greenhouse experiment on a sandy soil under temperate climate conditions, plant growth (and thus soil fertility) generally increased with increasing amendment of biochar and compost. This effect was more pronounced in a (nutrient-poor) sandy soil compared to a loamy soil. Biochar application rates were generally low with a maximum of 10 kg biochar per ton of compost material.

Fischer and Glaser (2012) also note that, additional amendments of clay minerals to biochar can promote CEC and Water Holding Capacity due to their high adsorption or swelling capacity. Furthermore, their incorporation of a range of minerals into organic substrates promotes the formation of organo-mineral complexes initiated by the biological activity of soil fauna after subsequent soil application. This aspect seems important since SOM in *terra preta* is stabilized by interaction with soil minerals (Glaser et al. 2003).

Other amendments like ash, excrements, worm juice or urine contribute to the nutrients stock of the final composting product and can enhance microbial activity by their nutrient supply (Glaser and Birk 2011). According to Arroyo-Kalin et al. (2009) and Woods (2003), ash may have been a significant input material into *terra preta*. Furthermore, for providing adequate moisture conditions during composting, urine can be added instead of water for preventing the dehydration of composting piles while adding nutrients at the same time. In Japan rice bran is added with the biochar to increase both rate of composting and to increase yields of the compost (Yoshizawa 2005).

A number of companies are starting to produce biochar compost mixtures. Carbon Gold Ltd. (UK) mixes wood charcoal with coir dust, seaweed extract, beneficial microorganisms and produces a composted product. Palaterra (www.palaterra.eu) has developed a process for mixing wood wastes and biodigester slurry with biochar and anaerobically digesting this mixture. It is a four stage process that takes only four weeks to complete: first the wood waste and biochar are soaked in the liquid digestate and mixed. Second, the mix undergoes a step of hot aerobic decomposition. The third stage is anaerobic lactic acid fermentation. Finally, the material is dried and bagged as a finished product.

Pyrolysis Partners in the USA (Koehler et al. 2012) developed a mix that comprises compost and an enhanced biochar. They carried out extensive trials in Florida with a pine bark and a bamboo biochar enhanced with heat treated minerals, horse manure and clay at application rates of 380 and 1100 kg/ha. They also carried out trials with compost added to these enhanced biochars (see Table 4). There was a marked improvement in yield of both marketable and total fruit when pine-bark biochar or pine-bark EB was applied (Fig. 9). The highest total fruit came from a treatment with small application of compost and pine bark EB, although the fact that there were only two replicates of this treatment makes this inconclusive. It is possible that increase in yield could be due to complex biological and chemical changes within the rhizosphere resulting in hormetic effects and induced systemic resistance to pathogens as discussed in earlier sections.

Table 4. Treatments. Fertilizer was applied at full rate, half rate, and no rate.

Treatments		
Full Fertiliser	**Half Fertiliser**	**No Fertiliser**
No Biochar or EB[1] (Control)	No Biochar or EB	No Biochar or EB
15 g EB	15 g EB	15 g EB
5 g EB	5 g EB	5 g EB
15 g raw bamboo biochars	15 g raw bamboo biochar	15 g raw bamboo biochar
5 g raw bamboo biochars	NT[3]	NT
15 g raw ICM[2] pine biochars	NT	NT
15 g EB with only ICM pine biochars	NT	NT
5 g EB with only ICM pine biochar	NT	NT
15 g EB with only ICM pine biochar + 15 g compost	NT	NT
5 g EB with only ICM pine biochar + 5 g compost	NT	NT
Buffer area (same as Control, but not measured)		

Notes:
1. EB = Enhanced Biochar
2. ICM = ICM LLC who made the biochar at 600°C
3. NT = No Treatment

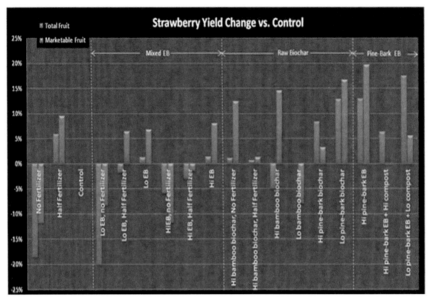

Figure 9. Strawberry yield changes as a function of soil treatments and fertilization rate. All blocks received the full complement of fertilizer except where labeled "No Fertilizer" or "Half Fertilizer." "Hi" and "Lo" refer to the treatment concentrations as detailed in Table 1. The error bars in these results are approximately ±5% for the Controls and mixed BMC data and ±10% for the raw biochars.

Color image of this figure appears in the color plate section at the end of the book.

5 Future Directions in the Development of Enhanced Biochars

A great deal more development work is required to be able to match biochars and biochar complexes to meet specific requirements of soil and plants. The chemical fertilizer industry has already developed a standard set of procedures to determine what specific soils require in terms of macro and micro nutrients. A similar approach can be taken with biochars.

Nutrients required by plants have been determined over many decades of research and these should be a guide for developing the biochar formulations for specific soils and specific plants. There are also recommended application rates and application times for NPK and micronutrients for individual crops once the properties of the soil have been measured. It is then important to understand the physical and chemical properties in the soil where the plants are to be grown to determine necessary biochar formulations that lead to both sustainable crop yields as well as improvements to soil health.

There are a number of different options for producing designer biochars. The first involves the mixing of different biochars to provide a range of properties. For example the farmer wants to grow tomatoes in a soil where overuse of chemical fertilizers and irrigation has resulted in higher levels of toxic organic and inorganic compounds and low pH. The soil may also lack a particular set of micro-nutrients. Fungal diseases may be an issue for the particular plant being grown. A possible formulation, therefore, could be a low temperature biochar from agricultural and animal residues to increase macro and micro nutrients, CEC, and resistance to fungal diseases, together with a higher temperature greenwaste biochar that has a high pH and high rates of adsorption of toxic compounds.

The second approach considers biochar as a "base" onto which inorganic or organic nutrients can be added. The surface concentration and the rate of release of the nutrients can be controlled by pre-processing or post-processing the biochars. Preprocessing can involve one or more of the following process steps:

a) Soaking the biomass in a solution containing an acid (phosphoric acid) or a base (KOH)
b) Soaking the biochar in a solution containing an ammonia salt, clay and other minerals
c) Surface activation with steam and air during the pyrolysis process
d) Finely grinding the biochar to increase surface area, where there may be a high concentration of radicals or dangling bonds
e) Chemically modifying the biochar with an acid or a base after pyrolysis (as mentioned previously, chemical modification using compounds with P, K and N can result in increased nutrient content)
f) Activating the surface of the biochar, adding a mixture of minerals and or chemicals containing NPK, and then baking this mixture on the surface of the biochar
g) Mixing the biochar with minerals and chemically modifying the surface structure through anaerobic or aerobic fermentation. To date there is insufficient data to determine what actually happens to different biochars when they react aerobically or anaerobically. Certainly there will be a series of complex redox reactions, mediated by microorganisms, which will result in the formation of surface organo-mineral complexes and organic molecules, as well as formation of new acids and bases (Gilbert and Banfield 2005).

Macro and micronutrients can be added after the biochar is made. Jarosiewicz et al. (2002) injected ammonia gas into hot biochar to increase the nitrogen content of the carbon lattice. Industrially this could be carried out in a separate reactor that is fitted to the end of a biochar reactor. Similarly a mixture of biomass with a high NPK content manure or digestor sludge and clay minerals could be added to the hot biochar as it leaves the reactor. This reactor could have a granulator or a pelleting press incorporated so that the material comes out as a uniform product. Baking these mixtures onto the surface has the advantages of sterilizing the biomass, cooling the biochar so it does not spontaneously combust, and controlling the rate of release of the nutrients. The work of Chia et al. (2012) indicates that there is a significant reaction between the biochar and the added organic and mineral matter. It is possible that coating biochar with nutrients results in plants having access to nutrients on demand as has been achieved by coating chemical fertilizers with polymers (Tomaszewska et al. 2002). Biochar can be added to chemical fertilizers to improve the nutrient use efficiency. Little work has been done in this area although early work in China indicates that this can significantly improve fertilizer efficiency (Pan et al. 2011).

Through preprocessing or post processing of biochars a fertilizer can be produced that best utilizes the numerous properties of biochar. Well resourced development and demonstration programs now need to be implemented both by commercial and research organizations working cooperatively.

References

Ahmadpour, A. and D.D. Do. 1997. Carbon 35: 1723–1732.

Amonette, J.E. and S. Joseph. 2009. Physical properties of biochar. pp. 33–53. *In:* J. Lehmann and S. Joseph [eds.]. Biochar for environmental management: Science and Technology. Earthscan, London.

Antal M.J. Jr. and M. Grønli. 2003. The art, science, and technology of charcoal production. Industrial and Engineering Chemistry Research 42: 1619–1640.

Arroyo-Kalin, Manuel, Eduardo G. Neves, and William I. Woods. 2009. Anthropogenic Earths of the Central Amazon Region: Remarks on their Evolution and Polygenetic Composition. In *Amazonian Dark Earths: Wim Sombroek's Vision*, edited by William I. Woods, Wenceslau G. Teixeira, Johannes Lehmann, Christoph Steiner, Antoinette WinklerPrins, and Lilian Rebellato, pp. 99–126. Springer. Berlin.

Azargohar, R. and A.K. Dalai. 2008. Steam and KOH activation of biochar: Experimental and modeling studies. Microporous and Mesoporous Materials 110: 413–421.

Bartlett, R.J. 1999. Characterizing soil redox behavior. pp. 371–395. *In:* D.L. Sparks [eds.]. Soil Physical Chemistry 2nd Edition. CRC Press.

Benjamin, T.J., G. Soto and J. Major. 2011. Biochar research final report. CATIE. Costa Rica.

Bird, B.R., D.W. Bird, B.F. Codding, C.H. Parker, and J.H. Jones. 2008. The "Fire Stick Farming" Hypothesis: Australian Aboriginal Foraging Strategies, Biodiversity and Anthropogenic Fire Mosaics. Proceedings of the National Academy of Sciences 105: 14796–14801.

Blackwell, P., S. Edgecombe, S. Joseph, P. Munroe, Y. Lin, C.H. Chia, L. van Zwieten, S. Kimber, and N. Foidl. 2009. Preliminary assessment of the agronomic value of Synthetic Terra Preta (STP). Asian Pacific Biochar Conference. www.anzbiochar.org/AP%20BioChar%20 Conference-may09.pdf.

Boehm, H.P. 2001. Carbon surface chemistry. pp. 141–178. *In:* P. Delhaes [eds.]. Graphite and precursors. CRC, Amsterdam.

Boonchan, S., M.L. Britz, and G.A. Stanley. 2000. Degradation and mineralization of high molecular-weight polycyclic Aromatic Hydrocarbons by defined fungal-bacterial cocultures. Applied and Environmental Microbiology 66: 1007–1019.

Bourke, J., M. Manley-Harris, C. Fushimi, K.A. Dowaki, T. Nunoura, and M.J. Antal. 2007. Do all carbonized charcoals have the same chemical structure? 2. A model of the chemical structure of carbonized charcoal. Industrial Engineering and Chemical Research 46: 5954–5967.

Brodowski, S., B. John, H. Flessa, and W. Amelung. 2006. Aggregate-occluded black carbon in soil. European Journal of Soil Science 57(4), 539–546. doi:10.1111/j.1365-2389.2006.00807.x.

Calvelo Pereira, K.J., M. Camps-Arbestain, R. Pardo Lorenzo, W. Aitkenhead, M. Hedley, F. Macías, J. Hindmarsh, and J.A. Maciá-Agulló. 2011. Contribution to characterisation of biochar to estimate the labile fraction of carbon. Organic Geochemistry 42: 1331–1342.

Carter, S. and S. Shackley. 2011. Biochar Stoves: an innovation studies perspective UK Biochar Research Centre (UKBRC), School of GeoSciences, University of Edinburgh.

Chan, K.Y., L.van Zwieten, I. Meszaros, A. Downie, and S. Joseph. 2007. Agronomic values of green waste biochar as a soil amendment. Australian Journal of Soil Research 45: 629–634.

Chen, B., Z. Chen, and S. Lv. 2011. A novel magnetic biochar efficiently sorbs organic pollutants and phosphate. Bioresource Technology 102: 716–723.

Cheng, C.H., J. Lehmann, J.E. Thies, S.D. Burton, and M.H. Engelhard. 2006. Oxidation of black carbon by biotic and abiotic processes. Organic Geochemistry 37: 1477–1488.

Cheng, C-H, J. Lehmann, and M.H. Engelhard. 2008. Natural oxidation of black carbon in soils: changes in molecular form and surface charge along a climosequence. Geochimica et Cosmochimica Acta 72: 1598–1610. doi:10.1016/j.gca.2008.01.010.

Cheng, C-H. and J. Lehmann. 2009. Ageing of black carbon along a temperature gradient. Chemosphere 75: 1021–1027. doi:10.1016/j.chemosphere. 2009.01.045.

Chia, C.H., S. Joseph, P. Munroe, Y. Lin, A. Downie, S. Kimber, L. van Zweiten, A. Cowie, B.P. Singh, J. Lehmann, K. Hanley, and P. Blackwell. 2008. Development of Synthetic Terra Preta (STP): Characterization and initial research findings. Retrieved December 01 2009, from IBI official website: http://www.biochar-international.org/sites/default/files/Joseph_IBI_poster_PM.pdf.

Chia, C.H., P. Munroe, S. Joseph, and Y. Lin. 2011. Microscopic characterisation of synthetic Terra Preta. Soil Research 48: 593–605.

Chia, C.H., Z.M. Solaiman, S. Joseph, E.R. Graber, Y. Lin, J. Hook, P. Munroe, P. Blackwell, B.P. Singh, and S. Donne. 2012. The role of minerals and symbiotic biology in enhanced biochar performance at low application rates. Soil Research. Submitted.

Chia, C.H., W. Gong, S. Joseph, C.E. Marjo, P. Munroe, and A.M. Rich. 2012b. Imaging of mineral-enriched biochar by FTIR, Raman and SEM–EDX, Vib. Spectrosc. http://dx.doi.org/10.1016/j.vibspec.2012.06.006.

Contescu, A., M. Vass, C. Contescu, K. Putyera, and J.A. Schwarz. 1998. Acid buffering capacity of basic carbons revealed by their continuous pK distribution. Carbon 36: 247–258.

Dias, B.O., C.A. Silva, F.S. Higashikawa, A. Roig, and A. Sanchez-Monedero. 2010. Use of biochar as bulking agent for the composting of poultry manure: Effect on organic matter degradation and humification. Bioresource Technology 101: 1239–1246.

Dixon, K. 1998. Smoke Germination of Australian Plants RIRDC report (98/108, KPW-1A).

Doydora S.A. , M.L. Cabrera, K.C. Das, J.W. Gaskin, L.S. Sonon, and W.P. Miller. 2011. Release of Nitrogen and Phosphorus from Poultry Litter Amended with Acidified Biochar. Int. J. Environ. Res. Public Health 8: 1491–1502.

Downie, A., A. Crosky, and P. Munroe. 2009. Physical properties of biochar. pp. 13–32. *In:* J. Lehmann and S. Joseph. [eds.]. Biochar for environmental management: Science and technology. Earthscan, London.

Elad, Y., D. Rav David, Y. Meller Harel, M. Borenshtein, H. Ben Kalifa, A. Silber, and E.R. Graber. 2010. Induction of systemic resistance in plants by biochar, a soil-applied carbon sequestering agent. Phytopathology 100: 913–921.

Enders, A., K. Hanley, T. Whitman, S. Joseph, and J. Lehmann. 2012. Characterization of biochars to evaluate recalcitrance and agronomic performance. Bioresource Technology 114: 644–653.

Fairhead, J. and M. Leach. 2009. Amazonian Dark Earths in Africa? pp. 265–278. *In*: W.I. Woods et al. [eds]. Amazonian Dark Earths: Wim Sombroek's Vision. Springer Science and Business media B.V.

Feng, Y., Y. Xu, Y. Yu, Z. Xie, and X. Lin. 2012. Mechanisms of biochar decreasing methane emission from Chinese paddy soils. Soil Biology & Biochemistry 46: 80–88.

Fengel, D. 1966a. On the changes of the wood and its components within the temperature range up to 200°C—Part 1. Holz Roh-Werkst. 24: 9–14.

Fengel, D. 1966b. On the changes of the wood and its components within the temperature range up to 200°C—Part 2. Holz Roh-Werkst. 24: 98–109.

Fischer, D. and B. Glaser. 2012. Synergisms between Compost and Biochar for Sustainable Soil Amelioration. pp. 167–198. *In*: S. Kumar and A. Bharti [eds]. Management of Organic Waste. InTech.

Gilbert, B. and J.F. Banfield. 2005. 'Molecular-Scale Processes Involving Nanoparticulate Minerals in Biogeochemical Systems', Reviews in Mineralogy and Geochemistry 59: 109–155.

Glaser, B., E. Balashov, L. Haumaier, G. Guggenberger, and W. Zech. 2000. Black carbon in density fractions of anthropogenic soils of the Brazilian Amazon region. Organic Geochemistry 31: 669–678.

Glaser, B. and W.I. Woods. 2003. Molecular Archaeometric Methods and the Origin of Amazonian Dark Earths: Possibilities and Limitations. Presented at the 68th Annual Meeting of the Society for American Archaeology, 13 April 2003, Milwaukee, WI.

Glaser, B. and J.J. Birk. 2012. State of the scientific knowledge on properties and genesis of Anthropogenic Dark Earths in Central Amazonia (terra preta de Índio). Geochimica et Cosmochimica Acta, 82, p. 39–51.

Graber, E.R., Y. Meller-Harel, M. Kolton, E. Cytryn, A. Silber, D. Rav David, L. Tsechansky, M. Borenshtein, and Y. Elad. 2010. Biochar impact on development and productivity of pepper and tomato grown in fertigated soilless media. Plant and Soil 337: 481–496.

Hammes, K. and M.W.I. Schmidt. 2009. Changes of biochar in soil. pp. 169–178. *In:* J. Lehmann and S. Joseph. [eds.]. Biochar for environmental management: Science and technology. Earthscan, London.

Heller-Kallai, L. 1997. The Nature of Clay Volatiles and the Condensates and their Effect on their Environment. Journal of Thermal Analysis 50: 145–156.

Hietala, S., S. Maunu, F. Sundholm, S. Jämsä, and P. Viitaniemi. 2002. Structure of thermally modified wood studied by liquid state NMR measurements. Holzforschung 56: 522–528.

Hillel, D. 2004. 'Introduction to soil physics.' (Elsevier India Pty Ltd: New Delhi).

Hina, K., P. Bishop, M. Camps-Arbestain, R. Calvelo-Pereira, J.A. Maciá-Agulló, J. Hindmarsh, J.A. Hanly, F. Macías, and M. Hedley. 2010. Producing biochars with enhanced surface activity through alkaline pretreatment of feedstocks. Soil Research 48: 606–661.

Hofrichter, M., D. Ziegenhagen, S. Sorge, R. Ullrich, F. Bublitz, and W. Fritsche. 1999. Degradation of lignite (low-rank coal) by ligninolytic basidiomycetes and their manganese peroxidase system. Applied Microbiology and Biotechnology 52: 78–84.

Hayashi, R. 1990. "Effects of purified wood vinegar as soil amendment and leaf surface spray." In *The Research Report on the New Uses of Wood Charcoal and Wood Vinegar*, ed. Technical Research Association for Multiuse of Carbonized Materials (TRA), Tokyo 331–341.

Heller-Kallai, L. 1997. The nature of clay volatiles and condensates and the effect on their environment: A review. Journal of Thermal Analysis 50: 145–156.

Hua, L., W. Wu, Y. Liu, M.B. McBride, and Y. Chen. 2009. Reduction of nitrogen loss and Cu and Zn mobility during sludge composting with bamboo charcoal amendment Environmental Science and Pollution Research 16: 1–9.

Huber, S., A. Balz, M. Abert, and W. Pronk. 2011. Characterisation of aquatic humic and non-humic matter with size-exclusion chromatography—Organic Carbon Detection—Organic Nitrogen Detection (LC-OCD-OND). Water Research 45: 879–885.

Illn-Gomez, M.J., A. Garcia-Garcia, C.S.M. de Lecea, and A. Linares-Solano. 1996. Activated carbon from Spanish coal. 2. Chemical activation. Energy and Fuels 10: 1108–1114.

Ishigaki, K., H. Fujie, and K. Suzuki. 1990. The effect of the soil amendment materials with charcoal and wood vinegar on the growth of citrus, tea plant and vegetables. In *The Research Report on the New Uses of Wood Charcoal and Wood Vinegar*, ed. Technical Research Association for Multiuse of Carbonized Materials (TRA), Tokyo pp. 107–120 (in Japanese).

Ishihara, S. 1996. Recent trend of advanced carbon materials from wood charcoals. Mokuzai Gakkai Shi 42: 717–723.

Ishii, H., S. Matsubayashi, and A. Nagai. 1990. Effects of purified wood vinegar on the growth of crop plants. In *The Research Report on the New Uses of Wood Charcoal and Wood Vinegar*, ed. Technical Research Association for Multiuse of Carbonized Materials (TRA), Tokyo 343–362 (in Japanese).

Jarosiewicz, A. and M. Tomaszewska. 2003. Controlled-Release NPK Fertilizer Encapsulated by Polymeric Membranes. J. Agric. Food Chem. 51: 413–417.

Jindo, K., K. Suto, K. Matsumoto, C. García, T. Sonoki, and M.A. Sanchez-Monedero. 2012. Chemical and biochemical characterisation of biochar-blended composts prepared from poultry manure. Bioresource Technol. Epub ahead of print.

Jones, R. 1969. Fire-stick Farming. Australian Natural History 16: 224.

Joseph, S., C. Peacocke, J. Lehmann, and P. Munroe. 2009. Developing a Biochar Classification and Test Methods. pp. 107–126. *In:* J. Lehmann and S. Joseph [eds.]. Biochar for environmental management: Science and technology. Earthscan, London.

Joseph, S., M. Camps-Arbestian, S. Donne, P. Munroe, C.H. Chia, Y. Lin, and A. Ziolkowski. 2010. Does biochar lower the energy required for plants to take up nutrients by changing the redox potential in the rhizosphere? Available from: http://www.ibi2010.org/wp-content/uploads/Joseph_Redox.pdf.

Joseph, S.D., M. Camps-Arbestain, Y. Lin, P. Munroe, C.H. Chia, J. Hook, L. van Zwieten, S. Kimber, A. Cowie, B.P. Singh, J. Lehmann, N. Foidl, R.J. Smernik, and J.E. Amonette. 2010. An investigation into the reactions of biochar in soil. Australian Journal of Soil Research 48: 501–516.

Joseph, S. and Z. Solaiman. 2011. Effect of Artificially Aged Biochar (BMC) on the mycorrhizal colonization, plant growth, nutrient uptake and soil quality improvement, Asia Pacific Biochar conference Kyoto. Available from: http://biochar.jp/11_ArtificiallyAgedBiochar.pdf.

Joseph, S., L. van Zwieten, S. Kimber, P. Munroe, C.H. Chia, J. Hook, Y. Lin, and A. Ziolkowski. 2011b. Developing Biochars that can be Applied at Low Application Rates; Field Results. Asia Pacific Biochar Conference Kyoto. http://biochar.jp/sub1.html.

Joseph, S., D.D. Khoi, N.V. Hien, Mai Lan Anh, H.H. Nguyen, T.M. Hung, N.T. Yen, M.F. Thomsen, J. Lehmann, and C.H. Chia. 2012. North Vietnam Villages Lead the Way in the Use of Biochar; Building on an Indigenous Knowledge Base. Available from: www.biochar-international.org/profile/Stoves_in_Vietnam.

Jurewicz, K., K. Babel, A. Ziolkowski, H. Wachowska, and M. Kozlowski. 2002. Amoxidation of brown coals for supercapacitors. Fuel Process. Technol. 77: 191.

Keiluweit, M., P.S. Nico, M.G. Johnson, and M. Kleber. 2010. Dynamic molecular structure of plant biomass-derived black carbon (biochar). Environ. Sci. Technol. 44: 1247–1253.

Koehler, S., E. Gerhardt, and S. Joseph. 2012. Improving Yields of Strawberries Grown in South Florida Through Addition of Compost Biochar and Minerals http://www.biochar-international.org/biblio/author/2333.

Kleber, M., P. Sollins, and R. Sutton. 2007. A conceptual model of organo-mineral interactions in soils: self-assembly of organic molecular fragments into zonal structures on mineral surfaces. Biogeochemistry 85: 9–24. doi:10.1007/s10533-007-9103-5.

Kögel-Knabner, I., G. Guggenberger, M. Kleber, E. Kandeler, K. Kalbitz, S. Scheu, K. Eusterhues, and P. Leinweber. 2008. Organo-mineral associations in temperate soils: Integrating biology, mineralogy, and organic matter chemistry. Journal of Plant Nutrition and Soil Science 171: 61–82.

Kookana, R.S., A.K. Sarmah, L. van Zwieten, E. Krull, and B. Singh. 2011. Biochar application to soil: Agronomic and environmental benefits and unintended consequences. Adv. Agron. 112: 103–143.

Krull, E. 2010. Biochar from Bioenergy—more than just a waste-product, Australian Bioenergy Conference.

Kuzyakov, Y., I. Subbotina, H. Chen, I. Bogomolova, and X. Xu. 2009. Black carbon decomposition and incorporation into soil microbial biomass estimated by ^{14}C-labeling. Soil Biology and Biochemistry 41: 210–219.

Kwon, S. and J. Pignatello. 2005. Effect of natural organic substances on the surface and adsorptive properties of environmental black carbon (char): pseudo pore blockage by model lipid components and its implications for N_2-probed surface properties of natural sorbents. Environmental Science & Technology 39: 7932–7939.

Lagaly, G. 1984. Clay-organic interactions. Royal Society of London Philosophical Transactions Series A 311: 315–332.

Lehmann, J., B. Liang, D. Solomon, M. Lerotic, F. Luizão, F. Kinyangi, T. Schäfer, S. Wirick, and C. Jacobsen. 2005. Near-edge X-ray absorption fine structure (NEXAFS) spectroscopy for mapping nano-scale distribution of organic carbon. Global Biogeochemical Cycles 19, GB1013. doi:10.1029/2004GB002435.

Lehmann, J. and S. Joseph. 2009. Biochar for environmental management. Science and Technology. Earthscan, London.

Lehmann, J., M. Rillig, J. Thies, C.A. Masiello, W.C. Hockaday, and D. Crowley. 2011. Biochar effects on soil biota—a review. Soil Biol. Biochem. 43: 1812–1836.

Liang, B., J. Lehmann., D. Solomon, J. Kinyangi, J. Grossman, B. O'Neill, J.O. Skjemstad, J. Thies, F.J. Luizão, J. Petersen, and E.G. Neves. 2006. Black carbon increases cation exchange capacity in soils. Soil Science Society of America Journal 70: 1719–1730.

Liang, B., J. Lehmann, S.P. Sohi, J.E. Thies, B. O'Niell, L. Trujillo, J. Gaunt, D. Solomon, J. Grossman, E.G. Neves, and F.J. Luizão. 2010. Black carbon affects the cycling of non-black carbon in soil. Organic Geochemistry 41: 206–213. doi:10.1016/j.orggeochem.2009.09.007.

Light, M.E., M.I. Daws, and J. van Staden. 2009. Smoke-derived butenolide: Towards understanding its biological effects. South African Journal of Botany 75: 1–7.

Lima, I.M. and W.E. Marshall. 2005. Granular activated carbons from broiler manure: physical, chemical and adsorptive properties. Bioresource Technology 96: 699–706.

Lin, Y., P. Munroe, S. Joseph, L. van Zwieten, and S. Kimber. 2012. Nanoscale organo-mineral reactions of biochars in ferrosol: an investigation using microscopy Plant Soil 357: 369–380.

Lin, Y., P. Munroe, S. Joseph, R. Henderson, and A. Ziolkowski. 2012b. Water extractable organic carbon in untreated and chemical treated biochars. Chemosphere 87: 151–157.

Lozano-Castello, D., M.A. Lillo-Rodenas, D. Cazorla-Amoro, and A. Linares-Solano. 2001. Preparation of activated carbons from Spanish anthracite I. Activation by KOH. Carbon 39: 741–749.

Lyonga S.N. 1980. Some common farming systems in Cameroon, their influence on the use of organic matter and the effects of soil burning on maize yields and on soil properties. In: *Organic Recycling in Africa*. FAO Soils Bull. 43: 79–86. FAO, Rome.

Major, J., A. DiTommaso, J. Lehmann, and N.P.S. Falca. 2005. Weed dynamics on Amazonian Dark Earth and adjacent soils of Brazil. Agriculture, Ecosystems and Environment 111: 1–12.

Major, J., J. Lehmann, M. Rondon, and C. Goodale. 2009. Fate of soil-applied black carbon: downward migration, leaching and soil respiration. Global Change Biology 16: 1366–1379. doi:10.1111/j.1365–2486.2009.02044.x.

Marley, N.A., P. Bennett, D.R. Janecky, and J.S. Gaffney. 1989. Spectroscopic evidence for organic diacid complexation with dissolved silica in aqueous systems—I. Oxalic acid. Organic Geochemistry 14: 525–528.

Matusik, J., A. Gawel, E. Bielanska, W. Osuch, and K. Bahranowski. 2009. The effect of structural order on nanotubes derived from kaolin-group minerals. Clays and Clay minerals 57: 452–464.

Mayer, J., S. Scheid, and H. Oberholzer. 2008. How effective are 'effective microorganisms'? Results from an organic farming field experiment. 16th IFOAM Organic World Congress, Modena, Italy, June 16–20.

Mingjie, G.J. 2004. *Manual for Bamboo Charcoal Production and Utilization.* Nanjing, China: Bamboo Engineering Research Center, Nanjing Forestry University.

McLaughlin, H., P.S. Anderson, F.E. Shields, and T.B. Reed. 2009. All biochars are not created equal, and how to tell them apart. Proceedings, North American Biochar Conference, Boulder, Colorado, August 2009. www.biochar-international.org/sites/default/files/All- Biochars--Version2--Oct2009.pdf.

Namgay, T., Singh, B., and B.P. Singh. 2010. Influence of biochar application to soil on the availability of As, Cd, Cu, Pb, and Zn to maize (Zea mays L.). Aust. J. Soil Res. 48: 638–647.

Nocentini, C., G. Certini, H. Knicker, O. Francioso, and C. Rumpel. 2010. Nature and reactivity of charcoal produced and added to soil during wildfire are particle-size dependent. Organic Geochemistry 41: 682–689.

Novotny, E.H., E.R. deAzevedo, T.J. Bonagamba, T.J.F. Cunha, B.E. Madari, V.dM. Benites, and M.H.B. Hayes. 2007. Studies of the Compositions of Humic Acids from Amazonian Dark Earth Soils. Environ. Sci. Technol. 41: 400–405.

Nguyen, B. and J. Lehmann. 2009. Black carbon decomposition under varying water regimes. Organic Geochemistry 40: 846–853. doi:10.1016/j.orggeochem.2009.05.004.

Ogawa, M. and Y. Okimori. 2010. Pioneering works in biochar research, Japan. Australian Journal of Soil Research 48: 489–500.

Ogawa, M. and Y. Yambe. 1986. Effects of charcoal on VA mycorrhiza and root nodule formations of soybean: Studies on nodule formation and nitrogen fixation in legume crops. Bulletin of Green Energy Program Group II 8, Ministry of Agriculture, Forestry and Fisheries, Japan. pp. 108–134.

Okutsu, M., D. Hashimoto, K. Fujiyama, M. Nagayama, K. Oda, K. Taguchi, and K. Nakazutsumi. 1990. The effect of the soil amendment materials with charcoal and wood vinegar on the growth of rice plants, apple trees, and vegetables. In The Research Report on the New Uses of Wood Charcoal and Wood Vinegar. (Ed. Technical Research Association for Multiuse of Carbonized Materials (TRA)) pp. 121–131 (TRA: Tokyo) [in Japanese].

Osinamea, O. A. and F. Meppeb. 1999. Effects of different methods of plant residue management on soil properties and maize yield. Communications in Soil Science and Plant Analysis 30: 53–63.

Pan, G., Z. Lin, L. Li, A. Zhang, J. Zheng, and X. Zhang. 2011. Perspective on biomass carbon industrialization of organic waste from agriculture and rural areas in China. J. Agric. Sci. Tech. 13: 75–82, in Chinese.

Pignatello, J.J., S. Kwon, and Y. Lu. 2006. Effect of natural organic substances on the surface and adsorptive properties of environmental black carbon (char): attenuation of surface activity by humic and fulvic acids. Environmental Science & Technology 40: 7757–7763.

Rajkovich, S., A. Enders, K. Hanley, C. Hyland, A.R. Zimmerman, and J. Lehmann. 2011. Corn growth and nitrogen nutrition after additions of biochars with varying properties to a temperate soil. Biol. Fertil. Soils 48: 271–284.

Roth, C. 2011. Micro-gasification: Cooking with gas from biomass 97, GIZ.

Sano, H., E. Tatewaki, and T. Horio. 1990. Effects of the materials for greening with charcoal on the growth of herbaceous plants and trees. In 'The Research Report on the New Uses of Wood Charcoal and Wood Vinegar'. (Ed. Technical Research Association for Multiuse of Carbonized Materials (TRA)) pp. 155–165 (TRA: Tokyo) [in Japanese].

Schnitzer, M.I., C.M. Monreal, G. Jandl, and P. Leinweber. 2007. The conversion of chicken manure to biooil by fast pyrolysis II. Analysis of chicken manure, biooils, and char by curie-point pyrolysis-gas chromatography/mass spectrometry (Cp Py-GC/MS). Environ. Sci. Health B. 42: 79–95.

Shinogi, Y. 2004. Nutrient leaching from carbon products of sludge, ASAE/CSAE Annual International Meeting, Paper number 044063, Ottawa, Ontario, Canada.

Singh, B.P. and A.L. Cowie. 2008. A novel approach, using ^{13}C natural abundance, for measuring decomposition of biochars in soil. In Carbon and Nutrient Management in Agriculture, Fertilizer and Lime Research Centre Workshop Proceedings. L.D. Currie, L.J. Yates (Eds.). pp. 549. (Massey University: Palmerston North, New Zealand).

Singh, B., B.P. Singh, and A.L. Cowie. 2010. Characterisation and evaluation of biochars for their application as a soil amendment. Aust. J. Soil Res. 48: 516–525.

Smernik, R. 2009. Biochar and sorption of organic compounds. In 'Biochar for environmental management. Science and technology' (Eds. J. Lehmann, S. Joseph) pp. 289–296. (Earthscan: London).

Solomon, D., J. Lehmann, J. Thies, T. Schäfer, B. Liang, J. Kinyangi, E. Neves, J. Petersen, F. Luizão, and J. Skjemstad. 2007. Molecular signature and sources of biochemical recalcitrance of organic C in Amazonian Dark Earths. Geochimica et Cosmochimica Acta 71: 2285–2298.

Spokas, K. and D. Reicosky. 2009. Impacts of sixteen different biochars on soil greenhouse gas production. Annals of Environmental Science 3: 179–193.

Steiner, C. 2007. Slash and Char as Alternative to Slash and Burn: soil charcoal amendments maintain soil fertility and establish a carbon sink. PhD thesis, 2006 Bayreuth University. PUBL. Cuvillier Verlag, Gottingen.

Steiner, C., B. Glaser, W.G. Teixeira, J. Lehmann, W.E.H. Blum, and W. Zech. 2008. Nitrogen retention and plant uptake on a highly weathered central Amazonian Ferralsol amended with compost and charcoal. Journal of Plant Nutrition and Soil Science-Zeitschrift Fur Pflanzenernahrung Und Bodenkunde 171: 893–899.

Steiner, C., N. Melear, K. Harris, and K.C. Das. 2011. Biochar as bulking agent for poultry litter composting. Carbon Management 2: 227–230.

Tomaszewska, M. and A. Jarosiewicz. 2002. Use of polysulfone in controlled-release NPK fertilizer formulations. J. Agric. Food Chem. 50: 4634–4639.

Uchimiya, M., L.H. Wartelle, K.T. Klasson, A. Chanel, Fortier, and M.L. Isabel. 2011. Influence of Pyrolysis Temperature on Biochar Property and Function as a Heavy Metal Sorbent in Soil. J. Agric. Food Chem. 59: 2501–2510.

van Zwieten, L., S. Kimber, S. Morris, K.Y. Chan, A. Downie, J. Rust, S.D. Joseph, and A. Cowie. 2010. Effect of biochar from slow pyrolysis of papermill waste on agronomic performance and soil fertility. Plant and Soil 327: 235–246.

Verheijen, F.G.A., S. Jeffery, A.C. Bastos, M. van der Velde, and I. Diafas. 2010. *Biochar Application to Soils: A Critical Scientific Review of Effects on Soil Properties, Processes, and Functions*. EUR 24099 EN, Office for the Official Publications of the European Communities, Luxembourg, 149 pp.

Violante, A. and L. Gianfreda. 2000. The role of biomolecules in the formation and reactivity towards nutrient and organics of variable charge minerals and organominerals. In 'Soil Biochemistry'. pp. 207–270. Bollag, J. and G. Stotzky (Eds.) Marcel Dekker, New York.

Woods, W.I. 2003. Soils and sustainability in the prehistoric new world. pp. 143–157. *In:*B. Benzing and B. Hermann [eds]. Exploitation and Overexploitation in Societies Past and Present. Lit Verlag, Munster.

Yariv, S. and H. Cross. 2002. Organo-clay complexes and Interactions. Marcel Dekker, New York.

Yoshizawa, S., S. Tanaka, and M. Ohata. 2007. Proliferation of aerobic complex microorganisms during composting of rice bran by addition of biomass charcoal. 26. In Proceedings of the International Agrichar Conference, Terrigal NSW, Australia, May 2007.

Zachrisson, O., M.C. Nilsson, and D.A. Wardle. 1996. Key ecological function of charcoal from wildfire in the Boreal forest. Oikos 77: 10–19.

[1]School of Materials Science and Engineering, University of New South Wales, NSW 2052, Australia and Nanjing Agricultural University, Nanjing 210095-China;
E-mail: joey.stephen@gmail.com
[2]NSW Department of Primary Industries, Wollongbar NSW 2477, Australia.
[a]E-mail: lukas.van.zwieten@dpi.nsw.gov.au
[3]School of Materials Science and Engineering, University of New South Wales, NSW 2052, Australia.
[b]E-mail: c.chia@unsw.edu.au
[c]E-mail: p.munroe@unsw.edu.au
[d]E-mail:yun.lin@unsw.edu.au
[4]Mark Wainwright Analytical Centre, University of New South Wales, NSW 2052, Australia.
[e]E-mail: c.marjo@unsw.edu.au
[f]E-mail: J.Hook@unsw.edu.au
[5]School of Biotechnology and Biomolecular Sciences, University of New South Wales, NSW 2052, Australia.
[g]E-mail: t.thomas@unsw.edu.au
[h]E-mail: shaunson26@gmail.com
[6]Discipline of Chemistry, University of Newcastle, Callaghan, NSW 2308, Australia.
[7]Biochar Solutions, 73 Mt. Warning Rd Mt Warning 2484, Australia;
E-mail: potaylor@bigpond.com
*Corresponding author

7

A Comparison of Methods to Apply Biochar into Temperate Soils

Don Graves

Introduction

Nelson Bays Mycorrhizas, 17 Wilkie Street, Motueka, 7120, New Zealand;
E-mail: dgraves@ihug.co.nz

Introduction

The objectives of this chapter are:

i) To Evaluate methods of biochar application to assist, by understanding of existing 'best management' practices and expertise regarding biochar, soil and crop carbon recycling; and through adoption of 'appropriate technology' for small scale gardens or larger scale field soils and crops. This involves the inclusion of biochar into 'rhizospheres', i.e., undisturbed soil zones adjacent to plant roots, root exudates and associated populations of root-affected soil microorganisms; and the preparation of soils and seedbeds, to assist with crop establishment, and or the restoration, remediation or rehabilitation of degraded soils.

ii) To compare the potential risks, benefits and effectiveness of biochar and soil carbon sequestration by 'conventional' tillage and 'conservation' tillage methods.

iii) To develop and evaluate economically and ecologically viable 'no-tillage' methods worth further investigation to safely and accurately deliver biochar (and or soil amendments) into soils at specific soil depths and application rates.

iv) To compare energy usage of biochar application methods.

v) To discuss the possible practical and ecological implications of methods of application to soils, of biochar feedstock size and density, of nutrient amendments, or of the potential effects of 'designer biochar' materials and production methods.

The methods used for the manual or mechanical addition of biochar to soils of are likely to be derived from existing methods used for any of the three most common types of biological and nutrient materials applied to soils (i.e., seeds, organic waste-streams and soluble fertilisers). In order to prepare soils and seed beds to enable placement of seeds, fertilisers, biochar or other soil amendments, conventional and organic farmers and gardeners may use either 'conventional tillage' or 'conservation tillage' tools (Ritchie et al. 2000, Baker et al. 2006, Macgregor 2007, Ashworth et al. 2010). The central hypothesis of this chapter is that the ecological effectiveness of biochar on soil quality will be significantly affected by tillage methods which diminish or else enhance soil structure, soil organic matter levels, or soil biological populations.

Research into the effectiveness of various techniques for biochar application to field soils has only rarely been investigated or reviewed (Marris 2006, Woolf 2008, Blackwell et al. 2009, Blackwell et al. 2010, Atkinson et al. 2010, Major 2010, Cook and Sohi 2010, Solaiman et al. 2010, Verheijden et al. 2010, Bishop et al. 2012). The practice of applying biochar to soils in commercial agricultural and horticultural operations is only just commencing, consequently there are currently no widely accepted practical guidelines or best management practices (Major 2010). Nonetheless, as previously discussed by Blackwell et al. (2009):

> "The effectiveness of applications of composts, animal manures, or mineral fertilisers are known to vary significantly whether they are incorporated or surface applied, banded, or broadcast, and similar responses can be expected to the method of biochar application".

1 Appropriate Technologies—Matching Methods with Purposes to Apply Biochar in Soils

The potential effectiveness of proposed methods to place biochar, compost or other recycled organic matter into surface soils, will depend upon the intended purposes (Blackwell et al. 2009), in particular to:

i) Sequester soil carbon from the atmosphere.
ii) Improve or maintain health of crops, soils and waters.
   • Restore physically, chemically or biologically degraded land.
   • Decontaminate polluted soils.
   • Mitigate risks of soil derived eutrophication of water.
iii) Enhance energy efficiency and profitability of agricultural, horticultural or forestry crops.

1.1 High Biochar Application Rates for the Purpose of Maximising Soil Carbon Sinks

In view of the growing evidence on global climate change and an urgent need to reduce atmospheric CO_2 emissions or to remove carbon from the atmosphere and to sequester it in soils, it is quite understandable why biochar may seem to offer a stable carbon source and apparently safe option for soil carbon sequestration.

If the purpose of sequestering biochar is primarily the financial motivation of producing or selling 'carbon sequestration revenues' derived from maximisation of biochar quantities applied to soils, then there are many cheaper and much easier methods of achieving carbon sinks or sequestration, including burial of biochar in lakes or the ocean, or in trenches, deep pits or abandoned mine shafts.

1.2 High Biochar Application Rates for the Purposes of Treatment of Contaminated Soils

Large or 'one off' quantities of biochar inputs and intensive biochar incorporation methods are probably suitable for the restoration of severely degraded soils such as mine spoils (Harley 2010). Similarly, large quantities of biochar and energy intensive conventional tillage methods could be used to thoroughly mix polluted soils and biochar for the purpose of improving the treatment effectiveness of heavy metal (e.g., chromium) (Bolan et al. 2012), or soils contaminated by anthropogenic organic pollutants (Wang et al. 2012).

1.3 High Biochar Application Rates to Mitigate Risks of Eutrophication of Water

If the purpose of applying biochar is to mitigate risks of eutrophication of water, then large volumes of biochar could be applied in deep trenches adjacent to waterways or riparian zones for the purpose of intercepting or filtering labile nutrients from livestock effluent or fertiliser runoff.

1.4 High Biochar Application Rates for the Purposes of Improved Soil Quality

Two critical questions on biochar application rates and potential complications for subsequent research measurements of biochar stability or the effectiveness on crops and soils were asked by Lehmann et al. (2006): "What are the long-term productivity gains of a one-time and large application of biochar compared to annual and small applications"? And "Can biochar become a routine management option in agriculture"?

There are many further questions requiring answers regarding practical cost-benefit analyses of the mechanical energy required to incorporate and mix large 'one-off' biochar applications compared to routine biochar incorporation of small amounts. Further environmental cost-benefit analyses are required regarding potential detrimental environmental costs such as loss of soil 'ecological services' by existing soil biota resulting from high biochar application rates, energy intensive and repetitive soil disturbance practices, or prolonged vegetation clearance. Conversely there could be possible benefits or gains of ecosystem services in severely degraded soils where biochar soil conditioning effects could potentially introduce or improve soil resilience (Blanco and Lal 2010, Harley 2010, Rees 2010), enhance plant and soil microbial biodiversity, or assist to re-establish effective soil nutrient cycling and soil carbon sequestration processes (Allen 1991, Pietikainn et al. 2000, van der Heijden and Sanders 2002, van der Heijden et al. 2008, Smith and Read 2008, Ogawa and Okimori 2009, Bahn et al. 2009, Thies and Rillig 2009, Solaiman et al. 2010, Graves 2010, Graves 2011, Graves 2012, Klironomos et al. 2011).

For the purposes of defining what large 'one off' biochar application rates are, it is necessary to contrast the relatively substantial logistical difficulties of mechanical incorporation methods, compared to theoretical application rates obtained in small scale trials for assessment of potential detrimental effects on soil biota such as earthworms, or biochar application used for the purposes of amending soil pH, or soil bio-physical properties such as improved water-holding capacity or soil bulk density (Blackwell et al. 2009). Annual application rates of between 1 or 2 tonnes of biochar carbon per hectare are easily achievable and relatively much less energy intensive if incorporated with other food production practices. Many questions require further research into the economic affordability and environmental effectiveness of larger biochar application rates trialled to date including 10, 20, 30, 50, 65, and 120t of biochar C per hectare.

2 The Precautionary Principle—The Need to Know More Before Over Re-acting

The 'precautionary principle' or precautionary approach states that if an action or policy has a suspected risk of causing harm to the public or to the environment, in the absence of scientific consensus that the action or policy is harmful, the burden of proof that it is not harmful falls on those taking the action. The 'precautionary principle' has often been advocated for the widespread early adoption of novel agricultural methods, the introduction of novel organisms into new territories, widespread agrochemical use, or field use of genetic modification technologies. Therefore it should be no surprise that the precautionary principle regarding biochar use has been advocated by some proponents of organic production systems. The 'precautionary principle' was adopted in evaluation and certification standards of IFOAM (International Federation of Organic Agriculture Movements) http://www.ifoam.org/about_ifoam/standards/index.html, and recognises that soils are extremely complex ecosystems, and in the case of biochar use, contends that "only a handful of issues about biochar application and use are understood in any detail, most issues are known unknowns, and there are quite likely to be unknown unknowns" (Ho 2009, Merfield 2012). Further precautionary concerns raised by organic producers and certification organisations include recognition that the application of biochar is irreversible, the unknown effects from proposed use of large 'one off' application rates, and the 5 to 10 year time scale usually required to determine soil changes from one state to another. Additional identified issues for concern include pyrolysis by-products found in biochar including potential toxins in VOCs (volatile organic compounds) and PAHs (polycyclic aromatic hydrocarbons) (Merfield 2012).

There are numbers of conventional and organic agricultural practitioners, scientists and other members of the public who are biochar sceptics and have expressed their concerns that the co-production and utilisation of biochar and biogas could potentially come at the expense of benefits of soil Carbon sinks derived from composts, animal manures or crop residues (Ho 2009). Perhaps the sudden, recent and growing enthusiasm of some biochar advocates could be interpreted by sceptics as an implied superiority of biochar over traditional methods of compost or soil organic matter carbon sequestration? However, some of this biochar scepticism is also due to recent observations of increased resource competition for using crop residues to produce ethanol, methanol and biodiesel as a replacement for traditional methods to improve or maintain levels of soil organic matter. Despite some of the real or perceived possibilities of competition regarding Carbon resources, this author advocates that biochar application practices and policies are likely to be synergistic with more traditional land management practices that seek to improve soil health. Perhaps advocates of biochar application and the advocates of other soil carbon sequestration methods may actually share many aims, particularly in regard to issues of provenance to avoid the use of contaminated or unsuitable biochar feedstock, and the similarity of protocols necessary to ensure the quality and provenance of appropriate nutrient inputs for certified organic production systems (Kristiansen et al. 2006, Macgregor 2007, Merfield 2012) and International Biochar Initiative's Standardized product definition and product testing guidelines for biochar that is used in soil http://www. biochar-international.org/characterizationstandard.

There could be further mutual interests or shared aims between traditional methods of Carbon recycling, incorporating and accumulating crop residues in soils, and the use of biochar in production soils. In particular similarities of methods could exist in regard to how much and how often to apply these soil carbon and nutrient amendments. However, large quantities of biochar or nutrient inputs in crop or pasture soils are not necessarily a desirable aim, especially when it comes to using concentrated nutrients, or else when soil pH could be affected. Instead, the effects of 'little and often' input methods could provide sufficient resources to enhance soil quality or plant health. This prudential method is frequently the 'perceived wisdom' or accepted 'best practice' when it comes to using compost or manure inputs.

Individuals and communities include both biochar enthusiasts and biochar sceptics regarding the potential for positive effects or negative effects of applying of biochar to make darkened soils, and the potential for collaborative opportunities for local energy supply, social, ecological and soil 'resilience' (Harris and Hill 2007, Hopkins 2010a, Hopkins 2010b, Rees 2010, Graves 2012, Stuart 2012).

3 Biochar Research Foci—Cultural Heritage, and Historic Scientific Legacies of Research into Ancient Amazonian 'Terra Preta de Indio', or Novel Biochar Soils 'Terra Preta de Nova'

As a result of the relative abundance of charcoal in tropical Anthrosols compared to charcoal in temperate climate Anthrosols, research focussed on the long-term effects of charcoal amendments in field soils has by necessity mostly examined Terra Preta de Indio soils. There is very little published biochar research derived from oral or written sources of indigenous communities' knowledge of Amazonian, Meso-Amer-Indian peoples, or Pacific Ocean peoples' (Lehmann 2009a, Steiner et al. 2010). Scientists, farmers and gardeners are only beginning to awaken a consciousness of the oral-history of indigenous peoples' knowledge of human-made soils, traditional tillage and crop irrigation practices, or the environmental effects of using historically important and often 'time-tested' traditional soil management and cropping methods.

In the late 20th and early 21st century, published biochar research into temperate climate human-made soils has only begun to examine short-term effects of using a range of feedstock materials, varied biochar soil application rates, within varied temperature and hydrology systems (Kolb et al. 2008, Laird 2008, Laird 2009, Hammes and Schmidt 2009, Atkinson et al. 2010, Blackwell et al. 2010, Levine 2010). Currently there is very little research to determine the long-term environmental effects and effectiveness of biochar prehistorically applied in temperate soils (Furey 2006, Calvelo Pereira et al. 2012). Consequently there is practically no biochar research on long-term effects of biochar or charcoal-amended soils on pasture, fodder crops, vines, orchards or forestry (MacKenzie and DeLuca 2006). However, in many active tectonic zones, regional volcanic eruption debris events provide pockets of wildfire-derived charcoal that could provide biochar researchers with useful comparisons of long-term effects of buried charcoal in temperate soils (Wardle et al. 1998, Wardle 2002).

Despite an awareness of the on-going doubts of biochar sceptics, or possible as yet unidentified but nonetheless potentially important or controversial questions about biochar use or production, there are nonetheless growing numbers of gardeners and farmers who are keen to start making and or using biochar to amend to their soils, sequester carbon and make new soil qualities, i.e., 'Terra Preta Nova'. Therefore the need to determine 'best practice' methods for biochar application to soils should seek appropriate technologies and methods for both smaller-scale biochar application in home gardens as well as larger-scale appropriate technologies for biochar production and application in broad acre crop soils.

3.1 Biochar Application and Recycling of other Soil Organic Matter Sources

The focus of this chapter is methodological, or in other words 'how to' effectively and efficiently apply biochar into temperate soils. However, despite the many desirable attributes that applying biochar may impart on soil quality or plant health, it is not necessary or desirable to neglect or diminish the need to maintain or enhance methods for carbon sequestration of soil organic matter derived from decaying flora and fauna. It is not simply a matter of applying either biochar or recycled crop residues, or biochar instead of crop residues, but instead applying current best practices to conserve and apply both these carbon sources into soils.

The scientific method and likewise the legal justice process utilise rational sceptism to determine probabilities indicated by the presence of absence of intersecting lines of evidence. Science proceeds by gathering information piece by piece like constructing a picture in a jigsaw puzzle. Some of the pieces of data and part of the 'big picture' about the use and effectiveness of biochar as a soil amendment may include existing bodies of 'evidence-based' knowledge and 'best management' practices for sustainable land stewardship. Thus, this chapter will briefly describe why it could be necessary to take lessons learnt from years of scientific research and practical experience in soil conservation, conservation agriculture, management of existing soil biological populations, and conservation tillage techniques.

This chapter also briefly investigates apparent similarities of effects derived from the use of traditional (pre-mouldboard plough) agricultural and horticultural practices of indigenous societies, and soil management and agricultural practices of modern methods of conservation tillage, 'no-tillage' and 'no-dig' technologies that produce correspondingly enhanced soil quality effects.

3.2 Integration of Biochar Application into No-tillage and other Soil Carbon Enhancement Methods

In testimony before the U.S. House Select Committee on Energy Independence and Global Warming, Dr. Johannes Lehmann avoided unintended or confusing meanings on the subject of the application and use of biochar when he stated: "Biochar must be integrated into existing food production systems and not be an alternative to food production, make use of already developed best-management practices such as no-tillage or conservation agriculture, and, for efficiency, build on residue collection systems that are already in place" (Lehmann 2009b).

There are many farmers, gardeners and scientists, both conventional and organic, who already advocate or routinely use 'best management practices' that enhance the incorporation and accumulation of soil organic matter from crop wastes, animal wastes and composts (Gershuny and Martin 1992). If biochar application can become a routine management option in agriculture as was advocated by Lehmann (2009b), i.e., "integrated into existing food production systems", there could potentially be technical complications for biochar or carbon sequestration researchers who aim to question or measure biochar degradation rates or stability in 'Terra Preta Nova' soils. Regular biochar soil inputs (i.e., little and often) could provide sufficient inputs to enhance soil quality or plant health, however on-going monitoring of biochar by assessment of C^{14} decay is less accurate in circumstances where there is not one single biochar application date (Major 2010).

3.3 Comparing Objectives and Outcomes of 'Conventional Tillage' and 'Conservation Tillage'

What is meant by the term 'to till'? According to the Oxford English Dictionary, till[3] (verb) [with object] is defined as to prepare and cultivate (land) for crops, Origin: Old English tilian 'strive for, obtain by effort', of Germanic origin; related to Dutch telen 'produce, cultivate' and German zielen 'aim, strive', also ultimately to till[1]. The current sense dates from Middle English. For what purpose do farmers or gardeners 'till' soils? What outcomes do 'conventional tillage' practitioners 'aim' or 'strive' for? What are the environmental effects of using so much energy and mechanical effort to use 'conventional tillage' methods? What is meant by the terms 'conservation tillage, 'no-tillage' or 'direct seeding'? What are the possible or predictable environmental effects of using 'conventional tillage' or 'conservation tillage' practices to incorporate biochar into soils?

'Conventional tillage' practitioners use a combination of tools to invert ('turn over'), break up, 'soften' pulverise and modify soils 'until'[1] soil disturbance and vegetation clearance methods have created soil and seedbed conditions considered suitable for any particular seed size, e.g., potatoes, corn, lettuce or carrots. Large plant propagules such as potato tubers usually require 'conventional tillage' methods to loosen and prepare soils prior to sowing and 'moulding up' ridge and furrow rows during crop growth. In contrast, 'conservation tillage' or 'direct seeding' practitioners deal with intact soils that they contend are already in a suitable condition to provide a seedbed, without any prior requirement for additional preparation by mechanical soil disturbance or vegetation clearance. Any tillage practice which retains at least 30% of surface cover from previous crop residues as mulch can be called 'conservation tillage'. Conservation tillage methods predominantly use herbicides and retain crop residues as 'anchored' mulch. 'Direct seeding' methods are primarily used for sowing small seeds including most annual crops or pasture plants. Some conservation tillage practitioners may use a single mechanical operation to directly place seeds and fertilisers or biochar into undisturbed or minimally disturbed soils. Conventional tillage practitioners aim to maximise soil disturbance effects and remove or bury plant residues of crops and weeds. In contrast, modern 'conservation tillage' practitioners aim to minimise soil disturbance effects, and predominantly use herbicides and retain crop residues as 'anchored' mulch.

3.4 Four Important Issues for Tillage Practices to Encourage Reliable Seed Germination, Seedling Emergence and Establishment

1. Concerning vegetation residue management, conservation tillage retains previous crop residues as a surface mulch, and thus requires a reliable mechanical ability to place seeds into soils below the mulch, and simultaneously avoid blockage of seed drills. Conventional tillage practices bury or remove vegetation to produce bare soil surfaces, and thereby increases risks of soil erosion by wind and water.
2. Fertilizers should be placed close to seedlings, but not so close to risk osmotic burning of seedling shoots or roots.

---

[1]According to the Oxford English Dictionary, till[1] (preposition & conjunction) less formal way of saying until. Origin: Old English til, of Germanic origin; related to Old Norse til 'to', also ultimately to till[3] . In most contexts till and until have the same meaning and are interchangeable. The main difference is that till is generally considered to be the more informal of the two, and occurs less frequently than until in writing.

3. The seed depth has to be tightly controlled. Seeds have limited energy reserves to enable shoots to reach the soil surface and commence photosynthesis; seeds placed precisely at the same depth enable seedling shoots to emerge at about the same time.
4. The seed has to be properly covered, in order to minimise soil moisture loss in seedbeds and maintain a high relative humidity for optimal seed germination.

3.5 To Mulch or Not to Mulch?—To Till or Not to Till?

An increasing number of gardeners and farmers recognise the many potential benefits of the effective management of soil disturbance effects and the maintenance of plant rhizosphere and soil microbial functions. Many gardeners and farmers understand the potential benefits of soil conservation methods from using mulch including vegetation residues, cardboard or other synthetic soil surface cover as a physical protection against soil erosion. These soil conservation and conservation agriculture practices have resulted in an increased use of green manure and cover crop rotations becoming included in home and commercial production systems, including the inter-row spaces of vineyards (Ingels et al. 1998, Fig. 1). However, the practice of growing cover crops or green manures usually involves conventional tillage tools, including the use of a shovel, plough or rotary hoe to 'turn under' or 'turn in' these plants before they produce mature seed or become too woody. The accepted purpose of burying green manure crops, including legumes and brassicas, is to provide 'main crop' plants with essential nutrients such as nitrogen, and to increase levels of soil organic matter to assist soil biology, soil structure and soil moisture retention. During the vegetative growth stage of cover crops and green manures these plants also provides living mulch that protects soil surfaces against wind and water erosion risks.

Organic crop production systems mostly rely on 'conventional tillage' methods for weed control and seedbed preparation. Recently certified 'organic' and so-called 'no-tillage' practices have been developed using heavy rollers to crimp and kill annual green manures as mulch to control annual weeds (Fig. 2). However, this method could more accurately be described as a 'low-till' method (i.e., 'reduced tillage' or 'minimum-tillage'), as this crop planting strategy is still reliant on conventional tillage methods to establish the preceding green manure crop (Merfield 2008). The use of roller-crimper machinery needs to be accurately timed with flower development to cause maximal death or growth suppression of annual ground covers or green manure crops prior to 'direct seeding' operations.

Steam weeding and flame weeding are also suitable for organic no-tillage weed control of annual weeds and crop residues. Certified organic

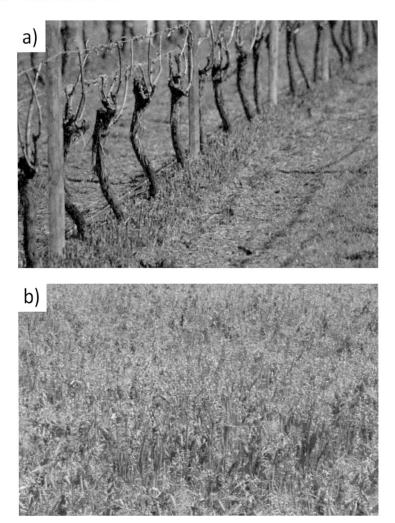

Figure 1. Examples of soil conservation methods: a) Grape Hyacinth groundcover growing in understory of vineyard. b) Peas and oats green manure crop.

Color image of this figure appears in the color plate section at the end of the book.

herbicides are desiccants, only suitable for killing annual weeds or green manures. Systemic herbicides that kill perennial weeds are not suitable for organic production systems. Suitable 'no-till' organic weeding methods are not foreseeable for controlling perennial weeds. The ability of no-tillage technologies to cope with crop residues or existing vegetation without operational disruption caused by machinery blockages is determined by the design of any particular seed drill. In New Zealand no-tillage methods

Figure 2. Roller-crimper used for non-chemical suppression of living mulches. Photos courtesy of Robert Merfield.

are commonly used to sow grass seed under pastures or sports turf, or the conversion of pine forest plantations to pasture. Some no-tillage technologies can be used to sow seed under crop residues as bulky as maize stems or else into pine forest 'slash' during 'trees to pasture' landscape vegetation conversion. The practice of 'heading', removing the inflorescence or the grass head during mechanical harvesting of cereal grass grains has increasingly left more plant stem length, which unless flattened can potentially contribute to risks of blocking or obstructing direct seed drills (Fig. 3). Effective weed control for no-tillage methods can be achieved by either 'blanket' sprayed application or else by banded herbicide application directly above the seedbed (Ritchie et al. 2000, Baker et al. 2006) (Fig. 4).

3.6 Mechanical versus Biological Biochar Incorporation in Soils

In order to address questions about how to best apply biochar into soils in a manner that will promote further mixing and incorporation by soil biological processes it is necessary to evaluate geographical and seasonal variations in biomass productivity that provide the food that soil biota consume, decompose and transport throughout the soil profile. Biochar soil amendments will enhance the total pool of soil organic carbon and may

Figure 3. No-tillage (direct seeding) into surface residues can occur immediately the previous crop has been harvested. Photo courtesy of C John Baker.

Figure 4. Side by side comparison of maize crops, direct seeded into blanket sprayed pasture (left) and conventional tillage sown seeds (right). Photo courtesy of C John Baker.

thus offer a complimentary source of soil water available for plants and soil microorganisms including arbuscular mycorrhizal fungi that confer drought tolerance effects to 'host' plants (Augé 2004).

Within the very broad category that includes all temperate soils there are many geographical, geological and climatic variations, and many dissimilar site specific land use histories that could complicate choices of suitable methods for the application of biochar to soils. Consequently there are currently many questions regarding how to maintain or enhance the ecological functions of soil microorganisms and biochar soil amendments.

3.7 Seeking Applicable Questions to Find Suitable Solutions for How to Apply Biochar

Thus, in recognition of the precept that unless we ask appropriate questions, we may not be seeking or else be able to find or recognise suitable answers (Kurlanski 2011), this chapter aims to ask if not yet fully answer some of the following questions:

i) Can we solve problems by using the same kind of thinking we used when we created them?
ii) If we always do what we've always done, will we always get what we've always got?
iii) What types of agricultural machinery should we use, or else aim to avoid, for incorporating biochar into soils?
iv) What are some of the potential environmental risks of using intensive methods of soil disturbance and vegetation clearance to apply biochar into soils?
v) How can biochar be applied into soils in a manner that reduces or mitigates unintended environmental risks?
vi) How can we best apply biochar into soils in a manner that will promote further mixing and incorporation by soil biological processes?
vii) What are the aims and outcomes of conventional tillage and conservation tillage methods?
viii) How could the proposed purposes and the resulting outcomes of applying biochar to soils potentially be affected by conventional tillage and conservation tillage methods?

3.8 Possible Answers or Potential Lessons from Social Sciences and Modern Soil Sciences

The answers to some if not all of the above proposed questions could possibly be located in examining charcoal-amended Anthrosols. However, in comparison to the large areas of tropical 'Terra Preta de Indio' soils, there are relatively very few examples of 'pre-historic' temperate climate Anthrosols formed by the intentional addition of charcoal (Best 1923, Furey 2006, Mitchell and Mitchell 2007). Thus, it may be difficult if not quite impossible to make many archaeological, anthropological or other scientific comparisons of charcoal-amended temperate Anthrosols and tropical Anthrosols. To say the least, it would be deeply regrettable if humankind has potentially lost so much cultural knowledge of traditional agriculture methods, including methods related to historic biochar production and use. Unfortunately we may no longer be able to ask for verbal or practical advice from the most knowledgeable and experienced practitioners of ancient charcoal production, or their traditional methods for the creation of dark coloured earths.

However, this paragraph briefly begins to explore how indigenous societies; in particular ngā iwi Māori and Polynesian societies, have traditionally practiced charcoal incorporation and conservation tillage practices. How could the cultural heritage of Polynesian or other traditional languages possibly assist modern societies to 'rediscover' or 'reconstruct' knowledge about the methods or aims of traditional agriculturalists and horticulturalists to create biochar or dark-coloured cropping soils?

Rudimentary linguistic analysis of words used to describe human-made or otherwise dark and humus-rich fertile crop soils, may disclose information about traditional Māori horticultural use of soil amendments; including man-made 'one paraumu'—'soil with charred wastes from earth-ovens', or else 'paraumu'—humus. Traditional Māori agriculture also incorporated the deliberate addition to soils of sand or gravel (Moorfield 2005). Could current day or future 21st century biochar researchers, gardeners and farmers possibly benefit from a deeper cultural understanding of the use of foot-plough tools including Māori digging sticks and subsequent secondary tillage methods of soil management? Could an 'in-depth' linguistic or cultural understanding of traditional Māori crop production and the use of modified soils hold helpful knowledge about similarities with digging stick methods used in indigenous Amazonian agricultural practices that may have helped to form 'Terra Preta de Indio' soils (Kramer 1966, Steiner et al. 2009)?

In the pre-European history of my country of origin, Aotearoa, New Zealand, indigenous Māori gardeners practiced a form of conservation tillage methods including 'raised-bed' gardens derived from their Polynesian ancestors. Traditional methods used 'digging sticks' or 'foot ploughs', known as 'Kō' or 'Kō whaka-ara' as primary tillage tools to break up soils (Fig. 5). This tillage method was followed by the use a number of hand tools and secondary tillage methods to further break clods apart in order

Figure 5. Entranceway carved figures of main historic and modern occupations of Motueka. Fisher (above), and Farmer depicted with traditional agricural tool. Image courtesy of master carver. Hone (John) Mutu and his assitant Tim Wraight of Te Awhina Marae, Motueka.

to form round mounds ('puke'), incorporate soil amendments (including sand, gravel and charcoal), and simultaneously to make pathways ('ara').

The use of foot-ploughs, Māori Kō or Scottish 'Cas Chrom' for primary tillage tasks, and the subsequent use of secondary tillage tools accommodated the preparation of contoured, rocky and rugged environment domains and steep slopes. The design of the 'Cas Chrom' is similar to a narrow spade with a long and 'crooked' handle for leverage. The method of using the 'Cas Chrom' was intended to 'turn over' a single slice of surface soil to make a narrow furrow (Best 1923, Kramer 1966) (Fig. 6).

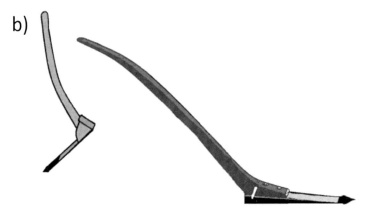

Figure 6. a) Cas Chrom ("crooked foot") or Scottish foot-plough being used in the Isle of Skye. Similar photographs from late 19th century Skye illustrate that women carried seaweed to be buried in the plough furrow. Photograph courtesy of Am Baile Highland Postcard Collection. b) Drawings of Cas Chrom ("crooked foot") Scottish foot plough.

Traditional Māori agricultural practices and the words used to describe traditional tillage tools of the Tūhoe nation were observed and language details recorded in early 20th century by a European ethnographer (Best 1923). These 'Euro-centric' but bilingual ethnographic observations recorded that traditional Māori tillage practices did not 'dig', 'turn over' or 'huri poki' soils (Moorfield 2005). The traditional tool names or words attributed to Tūhoe elders are descriptions of 'how to' use these tillage tools reveals that none of the words for traditional tillage tools can be interpreted as intended to be used to invert or 'huri poki' soils. Thus, a rudimentary linguistic te Reo Māori analysis of words recorded by (Best 1923) appears to confirm shared aims of traditional Māori agriculturalists and contemporary 'no-dig' gardeners or modern conservation tillage practitioners seeking to establish or renew pastures or to sow seeds of broad-acre crops without inverting topsoil (Best 1923, Graves 1994, Ritchie et al. 2000, Baker et al. 2006, Ashworth et al. 2010). In the relatively cold temperate climate of New Zealand, early Māori settlers successfully adapted to this unfamiliar and challenging climate using their traditional Polynesian tillage and raised bed methods in order to assist soil heat adsorption, improve drainage, assist moisture retention, and to minimise soil compaction effects. Traditional Māori gardeners deliberately modified soils as was necessary for the successful cultivation of imported tropical crops in cold temperate soils (Best 1923, Leach 1984, Furey 2006, Mitchell and Mitchell 2007). From the late 1830s many Māori agriculturalists had adapted quickly again and were early adopters of the recently perfected and commercialised 'mouldboard' plough technology. Māori communities subsequently established themselves as economically important farmers, crop growers, traders, exporters and shipping owners, provisioning large numbers of European whalers, sealers and settlers throughout Australasia and the sub-Antarctic. Traditional Māori garden methods with curvilinear rows of pathways and 'raised bed' round mounds ('puke') were quickly replaced by modern European 'soil inversion' methods with the characteristic straight lines of ridges and furrows formed by ploughs. Tropical-derived Māori crops grown in cooler temperate New Zealand soils included hue hue (gourd, *Lagenaria siceraria*). However, Māori mostly cultivated root crops including kūmara (sweet potato, *Ipomoea batatas* (L.) Lam), taro (*Colocasia esculenta*), and uwhi (yam, *Dioscorea* species) (Leach 1984, Mitchell and Mitchell 2007). In 21st century New Zealand most of these traditional crops are now grown almost exclusively in the warmer climatic conditions of the North Island, virtually no-one attempts to grow these tropical-derived root crops in the less favourable and cooler soils of the South Island. Human settlement and survival in diverse global climatic regions has often been characterised by an ability to adapt. Can contemporary 21st century New Zealand gardeners and farmers adapt their conventional land management practices to meet the increasing challenges of crop and soil conditions resulting from climate change?

In recent times Māori economic development researchers and primary production practitioners have been at the forefront of New Zealand institution-practitioner partnerships involved in developing and commercial use of mobile pyrolysis technologies (Knox 2012).

In 1840, a founding document of New Zealand, the Treaty of Waitangi was signed between the British Crown agents and traditional genealogy-based groups, iwi and hapū Māori. In 1975, the New Zealand Government acting as Crown partner of the Treaty instigated 'The Waitangi Tribunal', a permanent government commission of enquiry to hear evidence and recommend settlement terms for historic land confiscations and other Crown breaches including lost rights or risks to property rights guaranteed under the Treaty of Waitangi. The reports of the Waitangi Tribunal have determined that "the Treaty gives the Crown the right to govern, but in return requires the Crown to protect the tino rangatiratanga (full authority) of iwi and hapū Māori in relation to their 'taonga katoa' (all that they treasure)". The courts have characterised this exchange of rights and obligations as a partnership. One particular report into claim 'WAI-262' has highlighted the desire for consultation with Māori where such use of Māori treasures ('taonga') or Māori knowledge ('mātauranga Māori') is contemplated. The Waitangi Tribunal has also noted that "indigenous rights in cultural works were being debated internationally as part of global trade and IP processes". It also found that "international laws relating to IP do not constrain New Zealand from providing that protection. Reforms will allow New Zealand to become a global leader in indigenous rights, rather than a reluctant follower".[2]

3.9 An Historical Context of the Adoption and Decline of Conventional Tillage Methods and the Subsequent Growth of Modern Conservation Tillage Methods

The development of what is now called 'conventional tillage' was made possible in late 1700s by none less than the US president and agronomist Thomas Jefferson who was awarded in recognition of the mathematical perfection of the curved design of the mouldboard plough. Jefferson travelled widely and observed European advances in plough design that helped contribute to his design of the mouldboard plough. In the USA a practical plough design by Charles Newbold, of Burlington County, New Jersey, received a patent for a cast-iron plough in June 1797. However it was

---

[2]http://posttreatysettlements.org.nz/2011/07/waitangi-tribunal-report-on-wai-262-claim/

not until 1837 that John Deere developed manufactured and marketed the world's first 'self-polishing' cast steel plough. Early adopters and advocates of 'conventional tillage' nicknamed the mouldboard plough 'sod-buster' for its unique ability to convert ancient prairie grasslands into cropping soils. Farmers using mouldboard plough methods aim to bury plant 'wastes' and weeds, and have advocated soil inversion to bring 'fresh' minerals to the soil surface in order to establish the 'necessary' soil conditions for seedbeds.

Prior to the advent of modern environmentally 'benign' systemic herbicides the only scientifically defendable reason for ploughing was to kill weeds (Baker et al. 2006). The use and toxicity of glyphosate herbicide and surfactants remains a controversial topic for a number of sceptics, and is not permitted for use in certified organic production systems.[3] When glyphosate comes into contact with the soil, it can be rapidly bound to soil particles and be 'inactivated'. Further research is required to determine possible environmental effects of glyphosate adsorption by biochar. In wet soil conditions all tillage equipment can cause soil compaction effects. 'Steam plough' methods took over from horse drawn ploughs, and prior to the availability and popularity of tractor drawn ploughs were developed to avoid soil compaction risks associated with very heavy steam powered traction engines operating in field soils. In the late 19th century soil conservation methods to avoid soil compaction included use of the 'Fowler steam plough' technology, and were very effective at dredging lakes and ploughing very soft soils (Lane 1980, Brown 2008). Two huge traction engines were parked at far ends of a field and a relatively light weight double-ended 'Balance' plough or cultivator was attached to a wire rope and then winched back and forth between the two traction engines (Fig. 7).

The decline of the widespread use of ploughing, and the emergence and development of modern conservation tillage methods and 'no-tillage' technologies was instigated by publications of US agronomist (Faulkner 1945). Edward Faulkner challenged widely established agricultural practices and generations of land management, and blamed the universally used mouldboard plough for the disastrous "pillage" of the soil. Nature Magazine termed Faulkner's contentious new ideas as "an agricultural bombshell" when he stated "The truth is that nobody has ever exposed a scientific reason to till".[4] The widespread adoption of conservation tillage methods by farmers was not possible until effective technologies could be developed that could offer farmers dependable methods that are capable of providing

---

[3]In 2009 France's highest court ruled that US agrochemical giant Monsanto had not told the truth about the safety of its best-selling weed-killer, Roundup. The court confirmed an earlier judgment that Monsanto had falsely advertised its herbicide as "biodegradable" and claimed it "left the soil clean". http://news.bbc.co.uk/2/hi/europe/8308903.stm.

[4]http://www.ifoam.org/growing_organic/definitions/pioneers/edward_faulkner.php.

Figure 7. Paired 'Steam Plough' traction engines use wire ropes to tow a "Balance Plough" back and forth. Photos courtesy of Steam Plough Club, U.K.

equal or improved crop productivity, together with less environmental and economic risks than using conventional tillage practices. The growth in the global adoption of no-tillage technology increased from 8 million hectares in 1989 to 105 million hectares in 2009 (Derpsch and Friedrich

2009). In the latter half of 20th century post-colonial 'Western' academics and 'development' agencies began to recognise that social and applied sciences had a great deal to learn from each other, and furthermore began to acknowledge that two-way learning benefits could be possible between scientists from mostly industrialised nations and marginalised individuals from rural communities in poorer countries. As a result of this 'holistic' and interdisciplinary approach, the concept of 'appropriate technology' was derived from a recognition that either high-tech or low-tech methods could be 'fit for purpose', and that complex technologies that cost lots of money are not necessarily 'superior' or more 'fit' than 'simple' technologies that usually cost less or can be built locally with local resources. In modern 20th and 21st century academic interests in development studies, anthropology, rural sociology, and applied science studies have continued to analyse the intended and unintended ecological and economics effects of agricultural development and technology on recipient communities. Social and applied scientists identified where mistakes had occurred in promoting the transfer, adoption and use of introduced technologies that were not always the best 'fit' for the intended purpose of improving the survival of the poorest people working and feeding themselves from the lowest quality soils.

As a result of holistic and inter-disciplinary merger of 'development studies', anthropology and technology transfer programmes from western-based applied science, the concept of 'Participatory Technology Development' (PTD) was proposed by (Chambers 1983). The collaborative development of 'appropriate technology' methods has been adopted by many farmers and scientists concerned with understanding or achieving environmental, economic and social 'sustainability' of organic and conventional agriculture land use practices, often in 'marginal' soils, commonly focussed on maintaining and improving soil quality whilst producing subsistence crops or fodder crops in difficult climatic conditions (Kristiansen et al. 2006, Macgregor 2007). It is proposed that reducing the barriers to adoption of 'appropriate' methods to co-produce and apply biochar and 'syngas' pyrolyser technologies are likely to be derived from 'open-source' designs and collaborative farmer—scientist PTD projects. It was hypothesised by Joseph et al. (2012) that an open source pyrolyser design involving a number of Institutions and Practitioners' would:

1. lower the hurdle for biochar adoption
2. build capacity for pyrolysis and biochar production for research
3. design a unit based on collective wisdom
4. create felt ownership of a design

Thus it is also hypothesised that open source biochar soil application designs and methods involving a number of Institutions and Practitioners could likewise enable similar mutually beneficial outcomes for farmers, scientists and technologists.

4 Biochar Application using Conventional Tillage and Conservation Tillage Methods: Energy Costs and the Effective Management of Environmental and Occupational Risks

As previously stated, the central hypothesis of this chapter is that "the ecological effectiveness of biochar on soil quality will be significantly affected by tillage methods which diminish or else enhance soil structure, soil organic matter levels, or soil biological populations". Thus it anticipated that the development of safe occupational biochar handling methods to achieve environmentally safe and effective methods for application of biochar into temperate soils and rhizospheres appear likely to be derived from ecological analyses of soil resilience and 'post-carbon' sustainability management (Walker and Salt 2006, Heinberg and Lerch 2010, Hopkins 2010a, Hopkins 2010b, Rees 2010). In particular, future biochar application methods are likely to call on ecological analyses of how soil structure and soil biota may be influenced by modern methods of soil conservation (Seybold et al. 1999, Sparling et al. 2003, Blanco and Lal 2010); conservation agriculture (Clapperton et al. 2003, Harris and Hill 2007, Knowler and Bradshaw 2007); conservation tillage (Ritchie et al. 2000, Baker et al. 2006, Ashworth et al. 2010); and soil restoration ecology (Urbanaska et al. 1997, Waisel et al. 2002).

4.1 What is Wrong with Ploughing or Conventional Tillage Methods?

What is it that is inefficient or environmentally harmful about conventional tillage practices? Conventional tillage practices are not only energy intensive to achieve but are also expensive on soil carbon. Conventional tillage practices promote oxidation of organic carbon in soil into CO_2, thus contributing to the emission of this greenhouse gas. Conventional tillage practices take away much of the 'food-stuff' of the microorganisms that live in the soil and provide the biological mechanisms that promote soil structure and hold soil particles together. Soil microorganisms have direct effects on the soil health that the plants need to grow in which in turn maximises the total productivity of soil and plant ecosystems. Repetitive conventional tillage practices increase the threat of soil erosion and bit by bit gradually destroy the quality of valuable topsoil environment that plants need to grow in (Baker et al. 2006, Baker 2012). Thus, to re-iterate, intensive and repeated conventional tillage practices negatively affect soil microorganisms, destroy soil structure, diminish earthworm populations and their beneficial effects, and reduce levels of soil organic matter (Brady and Well 2002).

4.2 Why Apply Biochar into Rhizosphere Soils?

No-tillage methods used to apply biochar into the 'rhizosphere' or 'root zone' of crop soils are analogous to methods used to optimise the effectiveness of mycorrhizal fungal inoculation of plant roots (i.e., placement of a soil amendment(s), seed, biochar, bacterial or mycorrhizal fungal inocula, in close proximity to plant roots (Bethlenfalvay and Linderman 1992, Brundrett et al. 1996, Abbott and Murphy 2003, Smith and Read 2008). Mycorrhizal responses to biochar were amongst the pioneering works in biochar research (Ogawa and Okimori 2009), and were reviewed by Warnock et al. (2007). If the intended purpose of applying biochar as a soil conditioner is to maintain or enhance soil quality and plant health, there are many possible reasons why it may be desirable to aim to place biochar in the rhizosphere, precisely where most soil organic matter is usually located in natural or non-anthropogenic soil ecosystems (Miller and Jastrow 2000, Fitter and Hay 2002, Ayers et al. 2009, Jones et al. 2009). Biochar placed in the rhizosphere can adsorb, capture and recycle otherwise leached and mobile growth-limiting nutrients, thus acting in a similar manner to slow-release fertiliser. The effectiveness of biochar placement in root zones may primarily be explained by the improved nutrient availability to plant roots and symbiotic rhizosphere soil micro-organisms. For detailed definitions on the variable nature of 'symbiosis' and two-species population interactions (Fig. 8) (Odum 1983, Odum and Barrett 2005, Graves 2003).

The unintended effects conventional tillage disturbance effects on the ecological functioning of arbuscular mycorrhizas are relatively well studied and documented. However there is relatively little research focussed on soil disturbance effects on populations of soil rhizosphere and mycorrhizosphere associated bacteria. We are only beginning to identify what bacteria exist with mycorrhizal fungi, let alone what symbiotic roles these bacteria may perform with plant roots or fungi. The ecological effectiveness of applying biochar into rhizosphere soils and associated soil microbes can also be explained using 'non-expert' farmer terminology (i.e., 'feeding the soil'), as is advocated by many 'organic' practitioners regarding inputs of decomposed and recycled organic matter carbon and mineral nutrients. Banded application of nutrient-rich biochar produced from effluent, or biochar enriched with nutrients placed into active rhizosphere soils appear likely to increase the availability of these nutrients to plant roots. However, it is well documented that the effects of increased plant phosphorous nutrition can decrease levels of plant root symbiosis with arbuscular mycorrhizal fungi (AMF) (Miller et al. 1995, Miller 2000, Smith and Read 2008).

The application of biochar into the rhizosphere could assist or enhance soil and plant health, as in addition to the above factors, biochar may act as a habitat for soil microbes and refugia from predation (Nishio 1996,

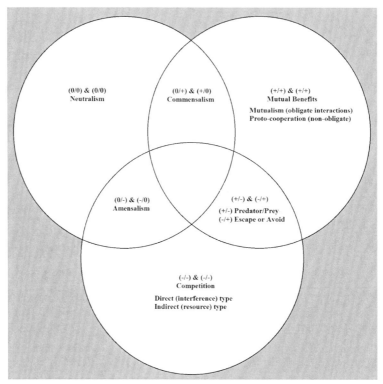

Figure 8. Venn diagram model of continuum of two-species population interactions (Graves 2003). Precautionary note: "The map is not the territory it represents, but if correct, it has similar structure to the territory, which accounts for it's usefulness" (Korzybski 1933), i.e., our views of reality are not reality itself but our own versions or understandings of it, our own mind's "map".

Pietikainen et al. 2000, Warnock et al. 2007, Ogawa and Okimori 2009); and could thus potentially assist an increase of soil microbial populations and diversity (Saito 1990, Saito and Marumoto 2002, Kolb et al. 2008, van der Heijden et al. 2008, Thies and Rillig 2009, Solaiman et al. 2010, Klironomos et al. 2011). However, conversely the ability of biochar to adsorb nutrients and organic compounds has been hypothesised to adsorb root exudates and thus may possibly interfere with plant-fungus signalling mechanisms that can assist the establishment of mycorrhizal symbiosis (Warnock et al. 2007). Further research is required to compare proposed plant-fungus signal interference effects where biochar has been thoroughly mixed with soils or else banded adjacent to or directly in rhizosphere soils (Solaiman et al. 2010). It has also been hypothesised that biochar applied to rhizosphere soils may also detoxify allelochemicals (i.e., toxic chemicals produced by a plant in order to defend itself against herbivores or competing plants), leading to altered root colonization by mycorrhizal fungi (Warnock et al. 2007).

The addition of low density biochar can also assist soil bulk density, and thus can enhance soil aeration and drainage associated with optimal soil structure. Increased levels of soil organic matter or biochar can assist soil water-holding capacity, thus mitigate or decrease potential drought risks, and decrease water stress effects in plants. In dryland soils, pastures and crops the application of biochar is likely to improve plant resilience and assist resistance to drought conditions. Banded biochar applied in close proximity to sown seed as well as 'side-dressed' application of biochar adjacent to established crop plant roots are likely to be the most effective methods to improve drought resilience of crops. The economic viability of applying biochar in soils is also dependant on mechanical and energy requirements to place biochar in shallow topsoil and rhizospheres compared to placement of biochar deeper in the soil profile.

Bio-geographical differences due to maritime and continental effects on temperate seasonal variations affect the continuity of vegetation growth. Under such a range of temperate climatic differences it could be anticipated that biochar applied within reach of plant roots will assist the water holding capacity of drought affected or 'marginal' soils and in turn increase biomass productivity and thus promote opportunities for further mixing and incorporation by soil biological processes including earthworms.

In a world where it is becoming increasingly difficult to predict anything with very much certainty, it may however be possible to predict an increased frequency of climatic extremes of temperatures or rainfall minima and maxima. Under both the increased occurrence of drought and floods it is probable that soil conservation matters will become more essential to increase protection against soil erosion by wind or water. Likewise it is probable that conservation agricultural methods will become increasingly important to maintain or improve levels of soil organic carbon that assist soil water holding capacity, aeration and drainage.

Whilst retention of crop residues and the use of cover crops or green manures are likely to be the predominant sources of plant derived organic matter, it could be a regrettable mistake to overlook or underestimate the ecological potential roles of weeds or non-crop plants for contributing to soil organic matter or as hosts of valuable soil micro-biota, especially AMF (Ocampo and Hayman 1981, Garcia-Garrido and Ocampo 2002). The more we understand the ecology of the rhizosphere and mycorrhizosphere the better equipped we are to manage the agroecology of soils in the future (Altieri 1995, Altieri and Nichols 2004). Predominant research and conventional crop management wisdom has it that weeds are strictly competitive with crop resources including water and nutrients. However, mycorrhizal research into the influence of 'non-host' crop rotations on arbuscular mycorrhizal colonisation also recognizes 'residual effects' on the extent of symbiotic mycorrhizal inocula present in cropping soils and 'host' plant roots (Ocampo and Hayman 1981). This research example offers a potential to recognise an ecological role of 'host' plant weeds as beneficial for maintaining populations of mycorrhizas, and thus reducing requirements for fertiliser inputs in following rotations of 'host' plant crops. Further research is required to investigate these residual effects of monocultures of 'non-host' crops and green manures including brassicas, beets, amaranths, buckwheat and lupins. Research is also required into hypothesised increased detrimental residual effects on arbuscular mycorrhizas from monocultures of herbicide-resistant 'non-host' crops, including 'Roundup-Ready' Canola (Graves 2003). The remediation of biologically degraded soils such as those following monocultures of 'non-host' crop rotations could in theory be achieved using expanded clay media (vermiculite, perlite or zeolite), or biochar particles used as a carrier medium for inoculating nursery propagation media, field soils and plants.

4.3 How Can Conservation Tillage Assist Soil Carbon Sequestration, Soil Structure, Mycorrhizal Functioning and other Soil Ecosystem Services?

In contrast to conventional tillage, conservation tillage methods have been shown to assist soil carbon sequestration and the subsequent formation and maintenance of physical stability of soil-aggregates (Bethlenfalvay and Linderman 1992, Oades 1993, Jastrow et al. 1998, 2007, Miller and Jastrow 1990, 1992, 2000, Simpson et al. 2004, Six et al. 2000, Six et al. 2002, Six et al. 2004, Rillig and Mummey 2006, Baker et al. 2006). Soil conservation and soil restoration ecological processes are significantly influenced by effective soil biological populations that are in turn influenced by rhizosphere nutrient availability (Blanco and Lal 2010). Soil organic carbon and hence available energy for soil biological processes may be derived from root exudates, symbioses with plant roots, and the metabolites of diverse populations of biotrophic or saprotrophic rhizosphere microorganisms including mycorrhizal fungi, bacteria and invertebrates (Ocampo and Hayman 1981, Urbanaska 1997, Coleman and Crossley Jr. 2004, Lal 2009, Smith and Read 2008). Ectomycorrhizal plants are obligate symbionts, whereas ectomycorrhizal fungi (Basidiomycetes and Ascomycetes) are non-obligate symbionts and may sometimes derive carbon from decomposed organic matter. Conversely, arbuscular mycorrhizal plants are often non-obligate symbionts and may 'thrive' on soluble nutrient inputs. However AMF are obligate biotrophic symbionts and cannot decompose organic matter, thus in the prolonged absence of suitable living plant hosts will die as they cannot grow independently, i.e., without the presence of living and compatible plant hosts. The functional ecology of AMF in temperate agricultural ecosystems has been shown to be influenced by the intensity of tillage. Furthermore, AMF are asexual, and thus do not have fruiting bodies such as mushrooms, puffballs or underground truffles that are capable of producing copious quantities of mobile and sexually produced spores. Intensive use of conventional tillage homogenises soils and dilutes or redistributes soil biological propagules including the asexual spores of AMF. The germination of AMF spores can occur independently, however the subsequent infection of 'compatible' plant roots is dependant the proximity of living plant roots and on the ability of spore AMF germination tubes to detect and move towards nearby roots.

The use of 'no-tillage' soil conservation methods has been shown to conserve soil 'biological fertility' (Douds Jr. et al. 2000, Douds Jr. and Jonhson 2003, Abbott and Murphy 2003, Clapperton et al. 2003). No-tillage methods do not destroy but instead maintain and assist the functioning of intact soil nutrient networks, including plant uptake of soil minerals and water enabled by symbiotic mycorrhizal soil fungal hyphae. Simultaneously intact networks of mycorrhizal hyphae also disperse plant-derived carbon into soils (Evans and Miller 1988, Bethlenfalvay and Linderman 1992, Helgason et al. 1998, Kabir et al. 1997, Kabir et al. 1998, Kabir et al. 1999, McGonigle et al. 1999, Miller 2000, Bahn et al. 2009).

Soil carbon sequestration rates are in part determined by vegetation history and vegetation habitats (e.g., deciduous broadleaf angiosperm forests versus conifer or gymnosperm forests); by plant mycorrhizal status (host versus non-host plants); by vegetation growth habits (e.g., annual versus perennial growth, monocots versus dicots); and branching surface roots versus deeper roots or taproot morphology (Waisel et al. 2002). Thus, it is proposed that the ecological effectiveness of methods to apply biochar to bare soils or biologically degraded soils could be optimised by revegetation methods and tillage techniques that improve rhizosphere nutrient availability.

Nutrient cycling differs significantly in temperate soils and tropical soils. In temperate soils organic matter decomposition rates are relatively slow and may result in sequestration of soil carbon and mineral nutrients, consequently most of the nutrients exist in the soil. In contrast, under the lush vegetation and high rainfall of tropical rainforest soils, relatively rapid rates of organic matter decomposition occur. As a consequence, in tropical rainforests most of the carbon and essential nutrients are locked up in living flora and fauna, in dead wood, and in decomposing leaves. As organic material decays, it is recycled so quickly that few nutrients stay long in soil, leaving tropical rainforest soils nutrient poor (Brady and Well 2002, Coleman and Crossley Jr. 2004). In temperate soil ecosystems, soil structure is enhanced by mycorrhizal fungi and other soil fungal hyphae. Soil fungal hyphae are 'thread-like' filaments that temporarily physically enmesh soil aggregates. Mycorrhizal fungal hyphae and associated soil zones, or 'mycorrhizospheres', support associated populations of soil bacteria that produce gums which assist chemical binding of soil minerals and soil organic matter to form physically stable soil micro-aggregates and soil macro-aggregates (Allen 1991, Bethlenfalvay and Linderman 1992, van der Heijden and Sanders 2002). Amongst other by-products of AMF is a 'glyco-protein' called 'glomalin' that glues soil particles into water-stable soil aggregates. This process of 'mycelia enmeshment', soil binding and aggregation is sometimes analogously described as a 'sticky string bag effect' (Miller and Jastrow 1990, Miller and Jastrow 1992, Miller and Jastrow 2000, Wright et al. 1999). Fungal derived carbon including glomalin is more resistant to decay than most carbon sources derived from plants, bacteria or animals, and thus fungal derived soil carbon has a relatively long residence time (Miller et al. 1995, Jastrow et al. 1998, Jastrow et al. 2007, Wright et al. 1999, Six et al. 2000, Six et al. 2002, Six et al. 2004, Simpson et al. 2004, Rillig and Mummey 2006, Fernandez et al. 2010).

Intensive and repeated soil disturbance can also decrease the frequency and physical resilience of water-stable soil aggregates, or soil 'crumbs'. The loss of physical resilience of soil aggregates can later cause them to collapse when wet, increasing soil compaction, and dramatically decreasing soil aeration, soil drainage, nitrate availability and crop yields (Alvarez and Steinbach 2009). Decreased levels of oxygen in soil water and soil air lessen oxygen available for respiration by plant roots and aerobic soil microbial populations. On a microscopic scale, intensive and repeated soil disturbance un-couples soil aggregation processes and affects the dynamics of soil carbon credit and debt of very small soil micro-habitats, including within soil 'crumbs'. Aerobic root and soil microbial processes such as respiration can occur in micro-habitats on the outer regions of soil micro-aggregates, and in regions between soil micro-aggregates and macro-aggregates. In microbial food-webs inhabiting regions below these aerobic surface regions of soil micro-aggregates, obligate anaerobic soil microbial processes such as fermentation and nitrogen fixation are able to occur. The intensive use of 'conventional tillage' soil disturbance breaks open soil aggregates and can destroy or diminish obligate anaerobic microbial processes by exposure to oxygen (Coleman and Crossley Jr. 2004). The physically stability of biochar particles and the internal conditions within particles of biochar has been analogously compared to soil aggregates formed by the effects of soil microorganisms on organic matter and soil minerals (Lehmann et al. 2011). When viewed on the micrometre or nanometre scale, uncompressed biochar particles may have large internal surface areas, a range of micropore sizes, heterogeneously distributed mineral or ash inclusions, and potentially high relative humidity that could provide diverse and suitable micro-habitats for microbiological processes to occur (Miller and Jastrow 2000).

The atmospheric conditions within biochar 'granules' may be further analogous to soil microaggregates, providing a gradient of aerobic and oxidising conditions near the granule surface, and tending towards anaerobic and reducing conditions nearer the centre of the biochar particle. Soil organic carbon is relatively chemically stable in the predominantly anaerobic conditions found in the innermost zones of soil aggregates (Tisdall and Oades 1982).

4.4 Designs, Restraints and Capabilities of Different Conservation Tillage Technologies

There are now many technologies which can be called 'conservation tillage' and 'no-tillage' methods including 'precision seed drills' ('V' shaped) 'double disc' and 'triple disc' seed drills; ('U' shaped) hoe soil openers; ('⊥' shaped or 'inverted-T' shaped) soil openers; ('+' shaped, 'hybrid' disc and blade, e.g., '⅃ | Ꞁ' openers); as well as 'subsoilers' and 'mole ploughs'. The aims of 'ridge tillage' and dedicated wheel track agricultural practices are similar to aims of 'raised bed' gardeners, i.e., to assist soil conservation aims to avoid or minimise soil compaction effects within seed beds or rhizospheres. Ridge tillage methods can use either conventional tillage and conservation tillage tools. In the urban back yards of many late 20th and early 21st century gardeners, conservation tillage practitioners actively avoid the effects of 'turning over' soils caused by using a spade, shovel, garden fork or hoe, and instead prefer to use a variety of so-called 'no-dig' or 'no-dig, no-weed' methods (King 1946, Guest 1948, Fukuoka 1978, Poincelot and Bennett 1988, Dean 1994, Gillbert 2003, Dowding 2007, Stout 2011). 'No-dig' garden methods that reduce soil disturbance effects include the use of hand tools such as soil corers or hand trowels to excavate a minimal amount of soil in order to transplant seedlings directly into mulched and otherwise minimally disturbed topsoil (Fig. 9).

Research and practical possibilities may come from use larger soil corers or ergonomically improved adaptations of tobacco seedling hand transplanters (Fig. 10). 'No-dig' hand weeding methods can avoid the excessive removal of vegetation ground covers by the use of a kitchen vegetable knife or else use 'lawn weeder' hand tools to undercut selected plant roots.

In the applications to broad acre crops and pasture renovation, not all contemporary conservation tillage methods are equal in their effectiveness on producing quality seedbeds in variable soil types soil conditions, or vegetation cover. There are significant differences in the abilities of various conservation tillage technologies to prevent crop residues and 'anchored mulch' from accumulating, clogging and causing seed drill operational problems. There is also variation in the ability of various conservation tillage

Figure 9. (a) A bulb-planter or soil corer hand tool for transplanting seedlings in 'no-dig' gardens. (b) Biochar is proposed as a substitute or renewable supplement for peat-based potting media.

technologies to cope with geographic variation in soil density and uneven soil surfaces whilst still being able to deliver seeds, fertilisers or other soil amendments at uniform depths in the soil profile (Ritchie et al. 2000, Baker et al. 2006, Ashworth et al. 2010).

Figure 10. Tobacco 'replanter' for replacing seedlings into row gaps, can also manually dispense water and fertiliser in conventionally tilled cropping soils. Historic photo (right) courtesy of Motueka District Museum.

In a review of disc seeding in 'zero-till' farming systems (Ashworth et al. 2010), generally classified 'disc openers' into four design categories:

  i)   disc coulters (coulter blades can be flat or fluted with either a continuous or scalloped cutting edge);
  ii) double discs (including triple disc combinations);
  iii) single discs (with vertical or 'undercut' variants);
  iv) disc/tyne and disc/blade hybrids.

The biological effectiveness and possible unintended side effects of banded fertiliser application using no-tillage and 'reduced tillage' seed drills has been shown to vary according to the shapes of seed drills or soil openers, and the resulting shapes of seed beds (i.e., 'V', 'U', '⊥' or '+' shaped seedbeds), and whether the seed bed is also covered by soil (Baker et al. 2006) (Fig. 11).

The potential side effects of seed drill or soil opener shapes may include the following:

  • soil vapour loss, seed bed and seedling desiccation,
  • increased soil bulk density, soil compaction,
  • decreased soil drainage and soil oxygen diffusion (aeration),

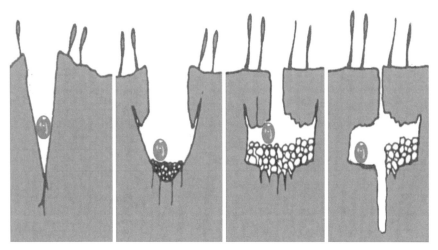

Figure 11. From left to right: "V"-shaped seedbed, formed by a "double-disc" seed drill; "U"-shaped seedbed, formed by a hoe seed drill; "⊥"-shaped seedbed, formed by a "Baker boot" seed drill; "┼"-shaped seed bed, formed by combination of disc and winged tine soil openers. Diagrams courtesy of C John Baker.

- burial and rotting of surface residue adjacent to seedlings, disease incidence,
- osmotic burning of seedling shoots and roots from concentrated fertilisers,
- soil surface runoff, wind or water erosion, contamination of surface waters (Ritchie et al. 2000, Baker et al. 2006).

Under moist soil conditions, or in 'sticky' or 'heavy' clay soils, almost all tillage methods and seed drills are prone to cause 'smearing' or soil compaction effects along the sub-surface soil contact zones with smooth metal surfaces of heavy farm machinery. The 'V' shaped seed bed is formed by a pair of discs (or 'double discs'), 'slanted' '\' and '/' to meet '\/' at the base. Double-disc '\/' and triple-disc '\ | /' low-tillage seed drills press downwards into the topsoil to produce a 'V' shaped and uncovered seed-bed, resulting in large soil vapour loss, as well as increased soil compression and oxygen diffusion effects in seedling root zones. Double-disc seed drills can be used in both undisturbed soils or in previously disturbed soils. The use of double-disc seed drills or 'V' shaped seed beds may also bury surface residues adjacent to seedlings and thus increase plant health risks from fungal pathogens derived from decaying crop residues. The double disc seeding method can also potentially cause osmotic burning of seedling shoots and roots derived from close contact with concentrated fertilisers. A 'U' shaped hoe or low-tillage seed drill produces a narrow seed-bed that may be covered with the topsoil, resulting in medium soil vapour loss, medium soil compression and oxygen diffusion effects. However, except where ground cover vegetation is sparse, the necessity to cover the seed with loose soil means that usually this seedbed preparation method cannot be used in undisturbed soils, thus the effects of soil disturbance and vegetation clearance is not significantly less disruptive to soil structure or soil biological nutrient cycling processes. The 'Baker Boot' seed drill produces an 'inverted T' shaped or '⊥' shaped seed bed with minimal soil vapour loss. The Baker Boot '⊥' shaped seed drill is often used in previously tilled soils, but can be used in undisturbed soils. The '⊥' shaped seed drill can be operated in conjunction with a vertically mounted metal disc to cut through crop residues preceding the seed drill opener. The '⊥' shaped seed bed is not usually closed by presser wheels following the opener. The '+' shaped seed bed is produced by a 'combination disc and winged tine soil opener' direct-seeding technology, purpose designed to avoid machinery 'blockage' problems commonly associated with operation in high volumes of crop residues. More specifically, the '+' shaped seed bed is produced by a combination of two horizontal slots (shelves) formed by paired 'ᒧ ' and ' ᒪ' shaped winged tine soil openers, and a vertical ' ‖' shaped slot formed by a single disc, i.e., the '+' seedbed shape results from the combination of 'ᒧ | ᒪ' soil opener tines and a disc. Finally, the '+' shaped seedbed is closed by a pair of offset angled compressor wheels following behind the soil opener. The ultra-low disturbance (ULD) effects of the '+' shaped seedbed design results in nil soil vapour loss, and minimal soil compaction or oxygen diffusion effects. Current design limitations of combination disc and winged tine openers allow for a maximum distance of 30mm vertical offset between the upper and lower ('ᒧ ' and ' ᒪ' shaped) openers and the resulting horizontal slot seedbeds (Ritchie et al. 2000, Baker et al. 2006, Ashworth et al. 2010).

Further to the soil opener shapes described above, it may be feasible to incorporate biochar with use of 'strip tillage' or 'zone tillage' methods, whereby a number miniature rotary hoes are used to 'power-till' narrow strips of soil (e.g., 10–15cm wide) ahead of (or together with) drill openers. The actual placement of seed and soil amendments varies with machinery design, but is either scattered or sown into a narrow zone of tilled soil. The soil between 'strip-tillage' sown rows remains undisturbed, thus 'zone tillage' can still be classified as 'conservation tillage' provided that 30% of surface residue ground coverage is retained (Baker et al. 2006). The establishment of broad acre crops together with direct application of biochar or other soil amendments will require well designed and large scale direct-seeding technologies. Practical design considerations include ease of loading and applying of liquid or solid fertilisers, different sizes and shapes of seeds and slurry particles. Further design fine-tuning will require optimal residue handling capability, and an ability across a range of soil types and soil moisture regimes to dependably produce seedbed shapes that optimise seed germination and seedling emergence equal to if not better than conventional tillage equipment abilities. Biochar application research and direct seeding peer-reviewed field research will need to be accompanied by practical field day trials allowing close evaluation by farmers, agronomy, and agricultural engineering researchers who can be seriously demanding and internationally competitive critics.

Conventional tillage methods are energy intensive compared to conservation tillage methods. Conservation tillage operations may be achieved with as few as one pass of a single tractor and a seed drill, whereas conventional tillage practices require many separate tractor drawn implements and many mechanical operations to invert and break up or pulverise soils in order to create a seed bed.

4.5 Describing Aims and Potentially Adverse Consequences of Producing 'Terra Preta Nova'

The terms 'Terra Preta de Indio' or 'Amazonian black earths' describe what can now be observed in the physical, biological and chemical states of these historic soils, and represents the product of countless generations of biological interactions and many years of hydrological reactions with these man-made soils. The reports of changes of biochar in soil and the virtually unrecognisable macrostructures of charcoal particles found in tropical 'Terra Preta de Indio' soils are possibly to be expected after countless interactions with many generations of earthworms, plant roots, soil fungi, and many other soil flora and fauna (Hammes and Schmidt 2009). The movement of biochar within soil strata may occur via the guts of various soil invertebrates, possibly following soil channels created by the death and decay of plant roots. Bacterial and fungal biochemical and enzyme mediated decomposition, together with bio-mechanical abrasion processes could also degrade the microscopic recognisable surface features of biochar.

Examinations of tropical 'Terra Preta de Indio' soils have shown that they sometimes contain pottery fragments, fish bones and other 'midden' wastes derived from kitchens and toilets. Thus, it appears likely that ancient Amazonian gardeners practiced a lot of organic waste recycling, not only charcoal. It also seems likely that the formation of 'Terra Preta Nova' by the mixing of biochar and other organic wastes prior to or during application to soils could potentially increase the density of plant roots in localised zones of nutrient enriched soils, and increase the palatability of biochar for ingestion by earthworms. The relatively nutrient rich, dark coloured and uniformly mixed characteristics of ancient 'Terra Preta de Indio' tropical topsoil layers could potentially provide modern biochar researchers with aspirational soil quality 'standards' or 'ends' to aim for when attempting to create modern biochar Anthrosols or 'Terra Preta Nova'. However, there could also be many unintended ecological consequences if modern biochar researchers and conventional tillage practitioners are tempted to choose to use intensive soil mixing machinery or any available technological 'means' in order to produce similar uniformly mixed dark coloured soils as quickly as possible. The unintended harmful consequences of intensive tillage on soil biota and soil structure are likely to be repeated if we choose to use intensive mechanical methods to instantly achieve a purpose of creating a visible appearance of uniformly dark 'Terra Preta Nova.'

4.6 Microscopic, Physical, and Biophysical Descriptions of Biochar

Biochar has a number of unique biophysical characteristics. The biological and cellular structures of the organic materials used as biochar feedstock are preserved during pyrolysis, and could assist soil biological fertility by providing suitable micro-habitats for soil microorganisms to occupy (Nishio 1996, Warnock et al. 2007, Kolb et al. 2008, Ogawa and Okimori 2009, Thies and Rillig 2009). The minute scale of pyrolysed biological structures provides a large surface area to volume ratio which consequently influences the ability of biochar surface charges to adsorb soil nutrients, contaminants and water (Hammes and Schmidt 2009). Biochar chemistry also contributes to adsorption properties through carboxylic groups. Last but not least important for carbon sequestration and environmental management of carbon sinks is the stability or longevity of biochar in the soil (Kutsch et al. 2009, Lehmann et al. 2009).

Biochar also has several unique physical characteristics, in particular a very low density, and being relatively easily crushed (i.e., 'brittle', or 'crystalline-like'). As a result of the brittle physical characteristics of biochar, transport or handling of biochar may result in the formation of dense clouds of black dust, decreasing visibility for machinery operators, resulting in an associated potential for explosive fire hazards, and a potential for public health respiratory risks (Blackwell et al. 2009, Atkinson et al. 2010, Major 2010).

Unfortunately there is a potential for interdisciplinary miscommunication regarding terms to describe the physical qualities of biochar. Some physicists and engineers involved in pyrolysis technology and biochar research describe this brittle quality as "friable", however this term has a very different meaning to soil scientists and farmers, i.e., not fragile or brittle, but instead having a crush resistant crumb or aggregate structure (Brady and Well 2002).

5 Methods of Biochar Application in Temperate Soils—Case Studies and Proposed Possibilities

5.1 Conventional Tillage Biochar Application Methods

Currently conventional tillage methods of biochar application are the most common case studies for establishing field trials and in small scale experimental methods using planter bags or pots grown in green house trials.

Hand application of biochar is very labour intensive, and prolonged contact with airborne biochar dust may cause occupational health risks. In field-scale case studies of biochar application trials, biochar broadcasting has been trialled using lime, fertiliser or manure spreaders, followed by mechanical soil mixing by disc harrow, rotary hoe or by mouldboard plough plus harrow (Fig. 12). The transport and surface spread application of dry and dusty biochar has several associated risks including losses of biochar dust to wind (Blackwell et al. 2009, Atkinson et al. 2010, Major 2010). Biochar dust can be a significant and potentially explosive fire risk as well as a visual hazard to machinery operators and a potential public health respiratory risk. Methods proposed to lessen the intensity of risks of biochar dust fractions

Figure 12. Hand application or mechanical broadcasting of biochar onto soil surfaces requires incorporation by harrow, rotary hoe or plough. Reprinted with permission of the Rodale Institute©.

include dampening prior to or during biochar application. However, the hydrophilic nature and moisture holding capacity of biochar could cause complications to biochar application methods that rely upon weight based quantification of application rates (Major 2010).

Surface spreading or 'top-dressing' of biochar could possibly require the least amount energy and consequently the lowest economic cost compared to all other possible biochar application methods. However, subsequent conventional tillage methods for mechanical incorporation of biochar by shovel, rotary hoe, power harrow, or serial combinations of plough, disc, tine, power harrows, or by 'trench and fill' methods could qualify as the most energy intensive and least economically viable of biochar incorporation or mixing methods (Williams and Arnott 2010). Nonetheless conventional tillage tools that thoroughly mix soils and biochar could be used to achieve purpose of rapidly achieving a homogenous dark soil similar in appearance to ancient 'Terra Preta de Indio' soils. The purpose of mechanical incorporation and thorough mixing of soil and biochar could possibly be to achieve maximal physical contact and maximal chemical interactions between polluted soils and biochar.

There are potentially beneficial roles for biochar application to assist remediation and restoration of soils where degradation of ecosystem services has been affected by conventional tillage, intensive soil disturbance, prolonged or repeated vegetation clearance, or inappropriate and prolonged anaerobic soil storage practices. Case studies of such intensive mixing of biochar and soils analogous to conventional tillage methods of incorporation of biochar include treatment of heavy metal (chromium) toxic contaminated soils including sites associated with timber treatment (Bolan et al. 2012). Further case studies using intensive methods for mixing of biochar and soils include decontamination of soils polluted with Anthropogenic Organic Contaminants (AOCs). It has been proposed that further research of biochar as a decontamination treatment of AOCs is likely to include Persistent Organic Pollutants (POPs), e.g., dioxins, PAHs, PCBs, etc. Pharmaceuticals and Personal Care Products (PPCPs), and Endocrine Disruptor Compounds (EDCs); and Pesticides (Wang et al. 2012).

Another possible purpose for thorough mixing of soil and biochar could be to increase the water holding capacity, soil bulk density, aeration or drainage of physically and biologically degraded soils. Biochar amendments could possibly assist in the remediation or restoration of soil ecology of mine spoil, stockpiled topsoil, or subsoil associated with large scale earthworks or landscape projects (Urbanaska et al. 1997).

A mouldboard plough was used to create a furrow, and a funnel directed biochar crumbs to the base of the furrow and plough zone (approx. ~15cm or ~6″ deep), directly behind the plough share (Fig. 13) (Bishop et al. 2012). The immediate purpose of this mouldboard plough method was to bury biochar

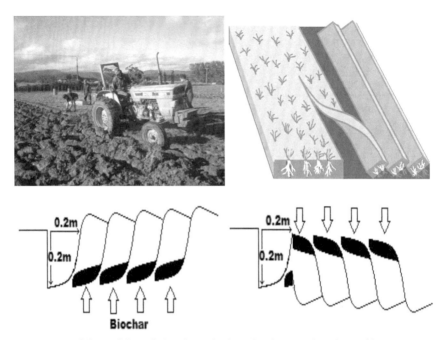

Figure 13. *Top left*: Mouldboard plough used to bury biochar crumbs at base of furrow. Photo courtesy of Bishop et al. (2012). *Top right*: Illustration of how a mouldboard plough is used to bury vegetation or biochar by inversion of soil surface layers. *Bottom left*: Illustration of how biochar was buried 20cm below the soil surface. Diagram courtesy of Bishop et al. (2012). *Bottom right*: Illustration of hypothesised surface location of previously buried biochar when disinterred by further soil inversion when re-ploughed.

where it can be protected from previously identified risks of biochar dust losses in windy conditions, or from surface water runoff losses. However, an unintended consequence of burying biochar at the base of the furrow could be that any future tillage practices on that trial site must aim to avoid further turning over and dis-internment of previously buried biochar. In effect such biochar burial methods could in turn demand 'no more tillage', or strictly only conservation tillage methods in order to protect against subsequent biochar dis-internment (Fig. 13, bottom right).

5.2 Biochar Applied in Trenches for the Purpose of Intercepting Labile Nutrients or Pollutants

Biochar could be applied in deep trenches for the purpose of intercepting or filtering labile nutrients such as livestock urine or nitrates where there could be environmental risks to surface water quality. Depending on the scale of a particular project, and or the location and soil types involved, this biochar application objective could be achieved manually or by mechanical diggers. A further and related purpose of this method could be to collect and re-apply the 'spent' or used biochar from such trenches or soil filters into crop or pasture soils as nutrient enriched and slow-release fertilisers.

5.3 Conservation Tillage Biochar Application Methods

Mole Plough and Subsoiler Biochar Application

As previously stated, the definition of 'conservation tillage' includes any practice that retains at least 30% of surface cover, thus the use of 'mole ploughs' and 'subsoiler' technologies can also be considered as conservation tillage methods. However, the purpose and resulting effects of mole ploughs and subsoiler techniques is not to create seedbeds, but instead to mitigate soil compaction effects by enabling soil drainage and aeration from the soil surface to depths often well below the reach of annual arable crop plant roots. Although biochar application trials using subsoiler methods have yet to commence, commercial research into these biochar slurry extrusion techniques has been discussed online by a company in Australia. The current size of most mole ploughs and subsoiler technologies could be modified to produce 'mini-mole plough' designs or similar adaptations of 'ultra-low disturbance' (ULD) direct seeder technologies. In New Zealand the results of preliminary research into the use of 'mole plough' techniques to apply biochar granules into cropping soils suggest that this method could also be applied to sequestration in the root zones of deep rooted perennial crops such as orchards, vineyards, or agroforestry systems (Bishop et al. 2012). These authors have speculated that deep placement of biochar may encourage roots to grow deeper to access soil water held in biochar. However, there is a considerable energy cost involved in the application of any machinery that is required to move through deeper soils. It is also difficult to separately measure or assess plant and soil biological effects caused by the addition of biochar, and the conjoined ecological and production effects caused by the mechanical mitigation of soil compaction, drainage and aeration.

Keyline Plough and Chisel Plough Biochar Application

Conservation tillage technologies that could be used to incorporate biochar include the use of 'chisel plough', and the 'keyline' plough aka 'Yeoman's Plough' (Hill 2002). The chisel plough is a primary tillage deep plough, working up to 50cm deep to ensure the development of a crop root systems at a depth below that usually achieved by a mouldboard plough. The chisel plough improves soil drainage and aeration and is ideal for use where top soil is fertile but subsoil is not so productive. 'Keyline planning' methods use the topographical form and shape of the land to determine the layout and position of farm dams, irrigation areas, roads, fences, farm buildings and tree lines. Contour ploughing by conventional tillage uses similar concepts to keyline planning to reduce soil erosion risks. 'Keyline plough' methods permit the storage of run-off water and soil organic matter on the farm itself and consequently effectively spreads the risks of irregular rainfall patterns common to dryland soils in rural production.

Deep Banded Biochar Application

Conservation tillage technologies also include the use of 'deep-banding' mechanical methods sometimes used to apply fertilisers or liquid slurry mixtures nearer the soil surface than achieved by 'mole plough' or 'subsoiler' technologies, but nonetheless below the root zones of most pasture plants, but within the root zones of some annual arable crops including sunflowers or maize. The application of biochar using 'deep banding' methods was achieved in previously tilled soils in Western Australia (Blackwell et al. 2009, Blackwell et al. 2010). The effects on aerobic soil biota of biochar placed below the rhizosphere of pasture roots is difficult to speculate, but indirect benefits to plants and soil biota could occur where forbs or rotations of other deep rooted plants may assist the upward translocation of soil water stored in biochar sequestered below roots of grass species.

Slurry Injector Biochar Application

The practice of using slurry injectors is not yet widespread, however an increasing number of agricultural contractors and farmers use slurry injectors to incorporate animal effluent associated with intensive production of cattle, pigs and poultry into pasture or crop soils. Current practices aim to reduce smell and recycle manure nutrients without contaminating surface water quality or degrading pasture palatability for livestock. Slurry injection methods may also be suitable for dairy shed effluent, poultry manure and biochar (Chan et al. 2008, Sohi et al. 2010). Biochar mixed with effluent may provide a possible method to reduce Nitrate pollution of groundwater, and decrease N_2O air pollution (Laird 2008, Laird 2009, Bruun et al. 2011). Slurry injectors can be generally classified into two design categories:

i) tractor or truck drawn tankers with combinations of injector discs and tines,

ii) tractor drawn discs and tines connected by an 'umbilical' flexible pipeline and pump to a supply tank at the edge of a field. In both design categories, sludge injector machinery must be able to function effectively on wet and potentially slippery soils whilst also minimising associated risks of soil compaction caused by heavy machinery wheel traffic on wet and plastic soils. Field trial research is required to optimise the transport and weight of biochar slurry manure or water combinations. It is not yet possible to determine optimal dilution ratios of liquid and biochar slurries or pastes. However, whether gravity fed, or transported by auger or by liquid pump, it appears likely that the drainage capabilities of the receiving soils together with rainfall and climate effects will determine the maximum slurry application rates, and the optimal travel speed of slurry injector machinery. Further research is required regarding practical and affordable methods to minimise risks of soil compaction associated with the use of heavy machinery and the application of biochar slurry. It could be of benefit to re-examine nineteenth century methods including the 'Fowler steam plough' and 'Balance Plough' technologies (Lane 1980, Brown 2008). Perhaps if field conditions are not congested by orchard or other trees, methods of application of biochar slurry could be adapted to use a similar winch system additionally including umbilical pipelines and pumps to apply slurry whilst minimising potential soil compaction risks? Perhaps a more practical arrangement could be to have liquid tankers located at either end of a field, with an open top tank in place of a fertiliser hopper on a direct drill, thus enabling ease of tank refills in order to optimize the weight required for a direct drill to be pushed downwards into soils, whist not unduly increasing risks of soil compaction to the receiving soils. As previously discussed, in order to increase the economic viability of such a proposed method, slurry could be applied adjacent to and perhaps slightly deeper than crop seeds.

Direct Seeder Biochar Slurry Application

The 'Baker Boot' direct seed drill has previously been trialled to assess its' effectiveness to apply liquid blood under sandy soil pastures. This experimental technique achieved the intended purpose to avoid surface contamination of pasture that would either be avoided by grazing livestock or could potentially pose animal health risks. Recently the author of this chapter conducted preliminary trial applications of 50:50 water diluted biochar slurry that was gravity fed into pasture soils using a single row combination vertical disc and winged tine soil openers and vertical disc direct seed drill. Using a low tractor speed this trial method was successfully able to apply a horizontal band of biochar slurry at a sub-surface seedbed depth of 4cm (Fig. 14). The liquid flow of biochar slurry produced unanticipated deposits of biochar slurry to a depth of 10cm, which was enabled by flowing down the vertical '$\parallel$' shaped slot produced by the seed drill disc. This field trial provided initial data for estimated biochar slurry application rates of approximately 1 litre of slurry per meter. At 0.15m spacing between seed drill openers, it would be possible to apply slurry at a rate of 6.1L/m^2, 61149L/ha, or 61.1m^2/ha. This slurry application rate is equivalent to 6.1mm rainfall.

Figure 14. Soil profile illustrating location of biochar slurry occupying both horizontal seedbed shelves and the verticle column produced by the disc designed to cut through crop residues (Graves unpublished data).

Further research regarding no-till biochar slurry application could be useful to investigate many issues, including sowing seeds of deep-rooted plants adjacent to vertical banded biochar. It could be potentially helpful to assess biotic interactions and possible movement of biochar resulting from the effects of plant life, death, decay cycles, and subsequent transport via earthworms using dead root channels (Graves unpublished data).

Direct Seeder Pelleted Biochar Application

Additional preliminary no-tillage biochar application trials were also recently conducted by this chapter author using combination disc and winged tine shaped soil openers, and scalloped edged disc technology to insert clay-coated biochar pellets at a subsurface depth of 4cm in pasture soils. This prototype method and novel biochar media was developed in order to test the practicality of using seed drills to apply dry biochar media whilst reducing potential ignition risks from biochar dust (Fig. 15) (Graves unpublished data). Clay-coated biochar pellets were hand scooped into the funnel as the tractor and seed drill moved at 0.188 meters per second. There were no significant difficulties experienced in applying the clay-coated pellets using the test seed drill apparatus. It was a relatively quick and

Figure 15. *Above*: Inspection view from above, clay-coated biochar pellets occupy one of two horizontal seedbed shelves located 4cm below the soil surface of an ungrazed pasture. *Below*: Soil profile illustrating clay-coated biochar pellets located in the leftmost horizontal seedbed produced by a single row combination disc and winged tine direct seed drill.

simple procedure to examine biochar pellets *in situ*, and to collect samples of clay-coated biochar pellets where they had been deposited by the seed drill. The biochar pellet application rate was estimated by collecting a sample of pellets from a 20cm length. The pellets sample was dried, then cleaned of soil and plant debris and weighed 20g. At 0.15m spacing between seed drill openers this corresponds to 100.4g/m, or 669.3g/m², or 6.69 tonnes of clay-coated pellets per hectare.

Following the apparent initial success of this trial, further research is sought to investigate the potential of clay-coated biochar pellets as a renewable carrier medium to introduce and inoculate biologically or physically degraded field soils and seedling plant roots with arbuscular mycorrhizal inocula (Fig. 16). Mycorrhizal inoculation of field soils has similarities to the anticipated benefits of adding biochar to soils, where nutrient poor soils are likely to be benefited more than nutrient rich or healthy soils (Graves 2010, Graves 2011).

Figure 16. Close up view of biochar feedstock *Pinus radiata* sawdust compressed pellets (left), charred pellets (centre), and clay-coated biochar pellets (right).

Color image of this figure appears in the color plate section at the end of the book.

Prospects for Direct Seeder Application of Mycorrhiza Inoculated Biochar Pellets

Current methods for mycorrhizal inoculation of plant seedlings usually involve using pot cultures or aeroponic methods to inoculate seedlings, followed by transplanting of seedlings into field soils (Brundrett et al. 1996, Smith and Read 2008). These mycorrhizal inoculation methods are most appropriate for assisting hand transplanting of high value mycorrhizal food crops such as blueberries or for revegetation and soil restoration projects. However, to date there have been relatively few available methods to achieve mycorrhizal inoculation of broad acre crops, or for establishing inoculated ground cover protection of erosion prone field soils. It is hoped that pot-culture methods and mycorrhiza inoculated coated or uncoated biochar pellets may offer an affordable and renewable substitute for expanded clay minerals such as vermiculite, perlite and zeolite that have previously been shown to be useful carrier media for on-farm production and utilisation of mycorrhizal inocula (Douds Jr. et al. 2010).

5.4 Small Scale Biochar Application by 'No-dig' Methods Including Transplanting, Dibber, Animal Manure and Seed Ball Methods

On the small-scale in home gardens or community gardens, biochar application could be achieved using previously discussed 'no-dig' transplanting methods including soil corers, trowels, dibbers, or digging sticks. Biochar application to garden soils or field soils could potentially include seedling transplants plus biochar potting media, plus or minus fertiliser or nutrient enhanced biochar, placed into the transplanting hole.

'No-dig' methods for biochar application could also include the broadcasting of 'seed balls', mixtures biochar, clay, topsoil or compost (Fukuoka 1978). The ecological application of biochar has been proposed via animal ingestion and deposits (or mechanical broadcast) of animal manures (Blackwell et al. 2009). Research into cover crops and no-till management for organic systems, using tractor drawn conservation tillage technology developed at Rodale Institute in Pennsylvania was the inspiration for New Zealand students who recently designed a prototype hand operated small-scale single row no-till seed drill for use in back yard gardens or in horticultural crops where 'small-scale organic growers could potentially 'catch up' with their broad-acre counterparts and explore the potential of no-till methods' (Guyton and Holmes 2012).

The inventors wish to increase the uptake of this small scale no-tillage technology by making it 'open-source'. The push-along single row hand-seeder nicknamed the 'Grasshopper' is similar in size to the conventional tillage 'Planet Junior' seed drill, but is capable of sowing plant seeds directly into lawn or through mulch without need for prior soil cultivation. The 'Grasshopper' no-till seed drill is fitted with a disc that cuts through soil surface mulch or ground cover to make a vertical slot, followed by a boot-shaped soil 'opener'. In theory the 'Grasshopper' design could potentially be an affordable and useful design for small-scale application of biochar in conventional or organic garden soils (Fig. 17). The 'Grasshopper' no-till seed drill design is similar to earlier small-scale prototypes developed in the 1980s by Massey University agricultural engineering researchers for trials

Figure 17. Small-scale "Grasshopper" no-till seed drill.

in Pakistan and Philippines. The earlier 1980s small-scale no-till prototypes were designed to be drawn by water buffalo or fitted to walk-behind hand tractors (Ribeiro et al. 2006).

6 Materials

Field soil testing of agricultural implement designs for no-tillage biochar application could possibly be assisted by movable soil bin methods to enable bench testing and climate control methods. The various purposes and proposed methods of biochar application will require different types of purpose designed biochar and this has led to the term 'designer biochar'. There are many types of biological feedstock that could be used to produce biochar for application to soils. There are also many possible variations of physical, chemical or biological biochar qualities that could be achieved by variations on pyrolysis methods. Purpose designed biochar, or 'designer biochar' could be focused on achieving a particular 'value added' feedstock nutrient content, shape, size, density, pelletizing, surface coating or soil microbial association. In order to assist biochar consumer or land users to visually verify the provenance of biochar feedstock it could be desirable to transport and store biochar in its original form or shapes prior to crushing into dust for soil application methods. The size and biochemical qualities of biochar materials are likely to affect its degree of attractiveness or palatability for ingestion, transport and nutrient benefits for soil invertebrates including earthworms. The degree of attractiveness or palatability is likely to be improved in a mixture of biochar and other organic matter amendments such as dairy effluent or compost.

Plant derived biochar feed-stock, nutrient and microbial soil amendments include the following choices:

- Large size wood pieces—tree trunks, root stumps derived from forestry, orchards or vineyards
- Medium size woody branches, shrub trunks, timber offcuts
- Small size woody twigs, bamboo stems, shoots, hops bines, grape vines, berry canes
- Very small chopped, shredded or whole stems, leaves and sawdust
- Compressed sawdust pellets
- Biochar dust in slurries or pastes

Further 'value-added' biochar including nutrient added biochar and feed stock choices could include:

- Food processing waste streams—fruit, skins, kernels, seeds, husks, fibre, fish and livestock meat offcuts, offal
- Livestock effluent—manure and urine
- Human toilet wastes, effluent or 'bio-solids'
- Spent or used charcoal filters
- Nutrient soaked sawdust, wood or bark chips
- Pelletised sawdust or bio-solids
- Clay-coated biochar pellets
- Biochar pellets containing clay, pyrolysis tar by-products, or sodium alginate
- Biochar pre-soaked in muddy water
- Biochar fragments as microbial carrier media and inocula sources

Acknowledgements

First of all my warmest and most sincere thanks are owed to my wife and best friend Maria, if it were not for your loving support none of this would have been possible.

Many thanks are also due to my family, friends and work colleagues for their encouragement to follow my passions to write, revise and re-write this chapter.

Enduring thanks and personal inspiration go to my mentor the late Dr A. Neil Macgregor for his outstanding example of commitment to life-long learning and to encouraging stewardship of healthy soils and soil biology.

I would like to pay special tribute to the hospitality and generosity of Tūhoe people for sharing and allowing the recording of their inherited knowledge of traditional agricultural methods and tools (Best 1923). Likewise I am personally very grateful to master carver Hone (John) Mutu, Tim Wraight, ngā whānau o Te Awhina Marae of Motueka for 'Manākitanga' (support and encouragement) to wider communities of speakers and teachers of Māori Language and Knowledge.

Special thanks (in no particular order) are due for the verbal support, helpful criticisms, questions, and generous practical assistance of Joanna and Jack Santa-Barbara, Jean-Paul Praat, Pari Rikihana, Colin Knox, Thierry Stokkermans, Clare Atkins, Sean DeLaney, Jason Duff, Gary Vergine, Michael Crofoot, David Douds Jr., Hailong Wang, Robert Hill, Shelly and Bruce Graves, Elza and Jehan Dissanake, Stephanie and Fraser Walls.

I am greatly appreciative to Natalia Ladygina and Francois Rineau for offering a unique opportunity to contribute to this text. I am also very thankful for valuable networking opportunities and many ideas shared at annual biochar workshops, and for the organising efforts of Marta Camps-Arbestain, Jim Jones and Linda Lowe of Massey University Palmerston North campus.

My personal respect and appreciation is due to the generous understanding and support of Marlene and Dr. C. John Baker, their personable phone discussions and helpful emails have been especially rewarding.

John's diverse work colleagues include farmers, scientists and manufacturers. John works relentlessly alongside others with shared aims to protect soil quality (pai Whenua), and our future food security. John inspires collaboration between scientists and land-users by having genuine respect for farmers' practical 'know how' or 'learn how to' adaptability, flexibility or resilience.

No reira, kia ora, tēnā koutou, kia ora koutou kātoa, many many thanks all!!

Personal appreciation is due for helpful assistance to obtain permission to reprint images, Charles (Merf) Merfield, Peter Bishop, Mike Henley, New Zealand Biochar Research Centre; Baker No-Tillage, C. John Baker, Rodale Institute, Adam Guyton, Hone Mutu, Tim Wraight, Tasman District Council, Steam Plough Club (UK), Motueka District Museum, Te Papa Tongarewa Museum Picture Library, 'Am Baile' Highland Postcard Collection, Scottish Library of Edinburgh Central Library, Paul Blackwell.

References

Abbott, L.K. and D.V. Murphy. What is Soil Biological Fertility? pp. 1–15, *In:* L.K. Abbott and D.V. Murphy [eds.]. 2003. Soil Biological Fertility—A Key to Sustainable Land Use in Agriculture. Kluwer Academic Publishers, The Netherlands.

Allen, M.F. 1991. The Ecology of Mycorrhizae. Cambridge Studies in Ecology. Published by Cambridge University Press, UK.

Altieri, M.A. 1995. Agroecology—The Science of Sustainable Agriculture. 2nd Edn. Published by Westview Press, Boulder, Colorado, USA.

Altieri, M.A. and C.I. Nichols. 2004. Biodiversity and Pest Management in Agroecosystems, 2nd Edn, Haworth Press, New York, USA.

Alvarez, R. and H.S. Steinbach. 2009. A review of the effects of tillage systems on soil physical properties, water content, nitrate availability and crop yields in the Argentinian Pampas. Soil Tillage Res. 104: 1–15.

Ashworth, M., J. Desbiolles, and E. Tola. 2010. Disc Seeding in Zero-Till Farming Systems—A Review of Technology and Paddock Issues, Published by Western Australia No-Tillage Farmers Association (WANTFA).

Atkinson, C.J., J.D. Fitzgerald, and N.E. Hipps. 2010. Potential Mechanisms for achieving agricultural benefits from biochar application to temperate soils: a review, Plant and Soil 337: 1–18.

Augé, R.M. 2004. Arbuscular mycorrhizae and soil/plant water relations. Can. J Soil Sci. 84: 373–381.

Ayers, E., D.H. Wall, and R.D. Bardgett. Trophic Interactions and Their Implications for Soil Carbon Fluxes, pp. 187–206 in Kuttsch M., W.L. Bahn, and A. Heinemeyer [eds.]. 2009. Soil Carbon Dynamics—An Integrated Methodology, Published by Cambridge University Press, UK.

Bahn, W.L., M. Kutsch, and A. Heinemeyer. 2009. Synthesis: Emerging Issues and Challenges for an Integrated Understanding of Soil Carbon Fluxes. pp. 257–271. *In:* L. Kuttsch, M. Bahn and A. Heinmeyer [eds.]. 2009. Soil Carbon Dynamics—An Integrated Methodology, Published by Cambridge University Press, UK.

Baker, C.J., K.E. Saxton, W.R. Ritchie, W.C.T. Chamen, D.C Reicosky, M.F.S. Ribero, S.E Justice, and P.R Hobbs. 2006. No-tillage Seeding in Conservation Agriculture (2nd Edn.) Publ. by CAB International and FAO.

Baker, C.J. 2012. Interview on Radio New Zealand National, http://www.radionz.co.nz/national/programmes/ninetonoon/audio/2520472/world-food-prize-nominee-dr-john-baker.asx.

Best, E. 1923. Māori Agriculture, published by NZ Government Print, Wellington, New Zealand.

Bethlenfalvay, G.J. and R.G. Linderman. 1992. Mycorrhizas in Sustainable Agriculture. ASA Special Publication Number 54, published by American Society of Agronomy Inc., Crop Science Society of America Inc., Soil Science society of America Inc., Madison, Wisconsin, USA.

Bishop, P., M. Hedley, C. Knox, P. Rikihana, and J. Jones. 2012. Practical Aspects on the Production of Biochar and its Delivery to Soil, New Zealand Biochar Research Centre, Proceedings of 2012 Biochar Workshop.

Blackwell, P., E. Krull, G. Butler, A. Herbert, and Z. Solaiman. 2010. Effect of banded biochar on dryland wheat production and fertilizer use in south-western Australia: an agronomic and economic perspective, in Aust. J. Soil Res. 48: 531–545.

Blackwell, P., G. Riethmuller, and M. Collins. 2009. Biochar Application to Soil, pp. 207–22. *In:* J. Lehmann and S. Joseph [eds.]. 2009. Biochar Environmental Management, Science and Technology.

Blanco, H. and R. Lal. 2010. Principles of Soil Conservation and Management. Published by Springer.

Bolan, N., J. Park, G. Choppala, S. Shenbagavalli, and S. Mahimairaja. 2012. Biochars Enhance the Remediation of Metal Contaminated Soils, presented at New Zealand Biochar Research Centre, Proceedings of 2012 Biochar Workshop.

Brady, N.C. and R.R Well. 2002. The Nature and Properties of Soils (13th Edition), Publ. by Prentice Hall, New Jersey, USA.

Brown, J. 2008. Steam on the Farm: A History of Agricultural Steam Engines 1800 to 1950. Published by Crosswood Press, Ramsbury, Marlborough, Wiltshire, UK.

Brundrett, M., N. Bougher, B. Dell, T. Grove, and N. Malajczuk. 1996. Working With Mycorrhizas in Forestry and Agriculture, Publ. by Australian Centre for International Agriculture Research (ACIAR) http://mycorrhizas.info/root.html#intro.

Bruun, E.W., D. Müller-Stöver, P. Ambus, and H. Hauggaard-Nielsen. 2011. Application of biochar to soil and N_2O emissions: potential effects of blending fast-pyrolysis biochar with anaerobically digested slurry. Eur. J. Soil Sci. 62: (4) 581–589.

Calvelo Pereira, R.M., M. Camps Arbestain, J. Vazquez Sueiro, J. Kaal, M. Sevilla, and J. Hindmarsh. 2012. Lessons From the Past: Carbon Fractions in Charcoal From Archaeological Māori Soils of New Zealand, presented at Biochar Workshop, Feb. 9–10, Massey University, Palmerston North, New Zealand.

Chambers, R. 1983. Rural Development—Putting The Last First, Published by Longman Press, UK.

Chan, K.Y., L. Van Zwieten, I. Meszaros, A. Downie, and S. Joseph. 2008. Using poultry litter biochars as soil amendments. Aus. J. Soil Res. 46: 437–444.

Clapperton, M.J., K.Y. Chan, and F.J Larney. Managing the Soil Habitat for Enhanced Biological Fertility, pp.1–15 in Abbott, L.K. and D.V. Murphy [eds.]. 2003. Soil Biological Fertility—A Key to Sustainable Land Use in Agriculture., Kluwer Academic Publishers, Netherlands.

Coleman, D.C. and D.A. Crossley Jr. 2004. Fundamentals of Soil Ecology (2nd Edition), Publ. by Elsevier Academic Press, Boston, USA.

Cook, J. and S. Sohi. Application to Soil, pp. 57–59. In: S.J. Shackley and S. Sohi [eds.]. 2010. An Assessment of the Benefits and Issues Associated with the Application of Biochar to Soil. http://www-test.geos.ed.uk/homes/sshackle/SP0576_final_report.pdf.

Dean, E. 1994. No-Dig Gardening and Leaves of Life, Harper Collins Publishing.

Derpsch, R. and T. Friedrich. 2009. Global Overview of Conservation Agriculture Adoption. 429–438 in Proceedings of the 4th World Congress on Conservation Agriculture, 4–7 February 2009, New Delhi, India. http://www.fao.org/ag/ca/CA-Publications/ISTRO%202009.pdf.

Douds, Jr., D.D., P.E. Pfeffer, and Y. Shacar-Hill. Carbon Partitioning, Cost, and Metabolism of Arbuscular Mycorrhizas. pp. 107–129. In: Y. Kapulnik [ed.]. 2000. Arbuscular Mycorrhizas: Physiology and Function, Kluwer Publications.

Douds Jr., D.D. and N.C. Johnson. 2003. Contributions of Arbuscular Mycorrhizas to Soil Biological Fertility. pp. 129–162 In: L.K. Abbott and D.V. Murphy. Soil Biological Fertility: A key to sustainable land use in agriculture, Kluwer Publications.

Douds Jr., D.D., G. Nagahashi, and P.R. Hepperly. 2010. On-farm production of inoculum of indigenous arbuscular mycorrhizal fungi and assessment of diluents of compost for inoculum production. Bioresource Technology 101: 2326–2330, or http://www.rodaleinstitute.org/node/441.

Dowding, C. 2007. Organic Gardening: The Natural No-Dig Way, Published by Green Books, UK.

Evans, D.G. and M.H. Miller. 1988. Vesicular–arbuscular mycorrhizas and the soil-disturbance-induced reduction of nutrient absorption in maize, I. Causal relations. New Phytol. 10: (1) 67–74.

Faulkner, E.H. 1945. The Plowman's Folly, Published by Michael Joseph Ltd, 26 BIoomsbury Street, London, W.C.1, UK.

Fernandez, R., A. Quiroga, C. Zoratti, and E. Noellemeyer. 2010. Carbon contents and respiration rates of aggregate size fractions under no-till and conventional tillage. Soil Tillage Res 109: pp. 103–109.

Fitter, A.H. and R.K.M. Hay. 2002. Environmental Physiology of Plants (3rd Edn.) Published by Academic Press, A Division of Harcourt Inc. Bodmin, Cornwall, UK.

Fukuoka, M. 1978. The One Straw Revolution—An Introduction to Natural Farming. Published by Rodale Press, USA.

Furey, L. 2006. Māori Gardening—An Archaeological Perspective Published by Science and Technical Publishing, Department of Conservation, Wellington, New Zealand. 137pages, http://www.doc.govt.nz/upload/documents/science-and-technical/sap235.pdf .

Garcia-Garrido, J.M. and J.A. Ocampo. 2002. Regulation of plant defence response in arbuscular mycorrhizal symbiosis, Review article in J. Exp. Bot. 53: (343) 1377–1386.

Gershuny, G. and D.L. Martin. 1992. Rodale Book of Composting—Easy Methods for every Gardener, published by Rodale Press, USA.

Gillbert, A. 2003. No-Dig Gardening. Published by ABC Books, Australia.

Graves, D.W. 1994. Kō Tillage, Soil Inversion in Perspective, in Proceedings of the 10th International Organic Agriculture (IFOAM) Conference, Lincoln University, Christchurch, New Zealand, December 11–16.

Graves, D.W. 2003. Models of Plant and Fungal Population Interactions During Mycorrhizal Symbiosis, presented at ICOM4, International Conference on Mycorrhizas, Montreal, Canada, and at ISS4 International Symbiosis Society, Halifax, Canada.

Graves, D.W. 2010. Biochar as a carrier medium to introduce fresh mycorrhizal inocula into plant root zones, presented at Biochar Workshop, Feb. 11–12, Massey University, Palmerston North, New Zealand.

Graves, D.W. 2011. Comparisons of No-tillage Technologies for Incorporating Biochar and Soil Micro-organisms Into Biologically Degraded Soils, presented at Biochar Workshop, Feb. 10–11, Massey University, Palmerston North, New Zealand.

Graves, D.W. 2012. Plant-Soil-Biochar Interactions on Ecological Resilience and Soil Resilience, presented at Biochar Workshop, Feb. 9–10, Massey University, Palmerston North, New Zealand.

Guest, A. 1948. Gardening Without Digging, Wigfield Publishers, UK.

Guyton, A. and N. Holmes. 2012. The Grasshopper, Organic NZ. 71: (4) 50–51, July–August, Published by Soil and Health Association on New Zealand Inc.

Hammes, K. and M.W.I. Schmidt. 2009. Changes of Biochar in Soil. pp. 169–181 *In*: J. Lehmann and S. Joseph [eds.]. Biochar Environmental Management, Science and Technology.

Harley, A. 2010. Biochar for reclamation in: The Role of Biochar in the Carbon Dynamics in Drastically Disturbed Soils. pp. 27–39 *In*: Levine [ed.]. U.S.-Focused Biochar Report: Assessment of Biochar's Benefits for the United States of America, Published by The Center for Energy and Environmental Security and The United States Biochar Initiative, http://www.biochar-us.org/pdf%20files/biochar_report_lowres.pdf.

Harris, A. and R. Hill. 2007. Carbon-Negative Primary Production: Role of Biocarbon and Challenges for Organics in Aotearoa/New Zealand, in Journal of Organic Systems 2: (2).

Heinberg, R. and D. Lerch. 2010. The Post Carbon Reader—Managing the 21st Century's Sustainability Crisis, Published by Watershed Media in collaboration with Post Carbon Institute, Santa Rosa, California, USA.

Helgason, T., T.J. Daniell, R. Husband, A.H. Fitter, and J.P.W. Young. 1998. Plowing up the Wood-Wide-Web, Nature 394: 431.

Hill, S. 2002. Yeomans Keyline Design for sustainable soil, water, agro-ecosystem and biodiversity conservation—a personal social ecology analysis in Agriculture for the Australian Environment. pp. 34–48 in proceedings of Fenner Conference on the Environment.

Ho, M-W. 2009. Beware the Biochar Initiative—Turning bioenergy crops into buried charcoal to sequester carbon does not work, and could plunge the earth into an oxygen crisis towards mass extinction. Institute of Science and Society, http://www.i-sis.org.uk/bewareTheBiocharInitiative.php.

Hopkins, R. 2010a. The Transition Handbook—From Oil Dependency to Local Resilience. Green Books, www.greenbooks.co.nz; http://transitie.be/userfiles//transition-handbook%281%29.pdf.

Hopkins, R. 2010b. What Can Communities Do? In The Post Carbon Reader—Managing the 21st Century's Sustainability Crisis, Published by Watershed Media in collaboration with Post Carbon Institute, Santa Rosa, California, USA, pp. 442–451.

Ingels, C.A., R.T. Bugg, G.T. McGourty, and I.P. Christensen. 1998. Cover Cropping in Vineyards —a Grower's Handbook, Published by University of California, Division of Agriculture and Natural Resources, Publication 3338.

Jastrow, J.D., R.M. Miller, and J. Lussenhop. 1998. Contributions of interacting biological mechanisms to soil aggregate stabilization in restored prairie. Soil Biol and Biochem. 30: 905–916.

Jastrow, J.D., J.E. Amonette, and V.L. Bailey. 2007. Mechanisms controlling soil carbon turnover and their potential application for enhancing carbon sequestration, Climatic Change 80: 5–23.

Jones, D.L., C. Nguyen, and R.D. Finlay. 2009. Carbon flow in the rhizosphere: carbon trading at the soil-root interface, Plant and Soil 321: 5–33.

Joseph, S., F. Christo, J. Lehmann, E. Fisher, S. Shabangu, A. Centre, and J. Jones. 2012. Development of an Open Source Pyrolyser, presented at Biochar Workshop, Feb. 9–10, Massey University, Palmerston North, New Zealand.

Kabir, Z., I.P. O'Halloran, J.W. Fyles, and C. Hamel. 1997. Seasonal changes of arbuscular mycorrhizal fungi as affected by tillage practices and fertilization: Hyphal density and mycorrhizal root colonisation. Plant and Soil 192: 285–293.

Kabir, Z., I.P. O'Halloran, J.W. Fyles, and C. Hamel. 1998. Dynamics of the mycorrhizal symbiosis of corn (*Zea mays* L.): effects of host physiology, tillage practice and fertilization on spatial distribution of extra-radical mycorrhizal hyphae in the field. Agriculture Ecosystems and Environment 68: 151–163.

Kabir, Z., I.P. O'Halloran, and C. Hamel. 1999. Combined effects of soil-disturbance and fallowing on plant and fungal components of mycorrhizal corn (*Zea mays* L.). Soil Biol Biochem. 31: 307–314.

King, F.C. 1946. Is Digging Necessary. New Times Publishers, UK.

Klironomos, J., M. Zobel, M. Tibbett, W.D. Stock, M.C. Rillig, J.L. Parrent, M. Moora, A.M. Koch, J.M. Facelli, E. Facelli, I.A. Dickie, and J.D. Bever. 2011. Forces that structure plant communities: quantifying the importance of the mycorrhizal symbiosis. in Forum/Letters New Phytol 189: 366–370.

Knowler, D. and B. Bradshaw. 2007. Farmers' adoption of conservation agriculture: A review and synthesis of recent research, in Food Policy 32: 25–48.

Knox, C. 2012. The Economics of Charcoal Production in Māori Owned Forests. Presented at Biochar Workshop, Feb. 9–10, Massey University, Palmerston North, New Zealand.

Kolb, S.E., K.J. Fermanich, and M.E. Dornbush. 2008. Effect of Charcoal Quantity on Microbial Biomass and Activity in Temperate Soils. Soil Sci. Am. J. 4: 1173–1181.

Korzybski, A. 1933. Science and Sanity: An introduction to Non-Aristotelian Systems and General Semantics, 5th Edition, published by Institute of General Semantics.

Kramer, F. 1966. Breaking Ground: Notes on the Distributions of Some Simple Tillage Tools, publ. by (Sacramento Anthropological Society Paper 5.) Sacramento, California, USA.

Kristiansen, P., A. Taji, and R. Reganold. 2006. Organic Agriculture—A Global Perspective, CSIRO Publishing, Australia.

Kutsch, W.L., M. Bahn, and A. Heinemeyer. 2009. Soil Carbon Dynamics—An Integrated Methodology, Published by Cambridge University Press, UK.

Laird, D.A. 2008. The Charcoal Vision: A Win-Win-Win Scenario for Simultaneously Producing Bioenergy, Permanently Sequestering Carbon, while Improving Soil and Water Quality. Agronomy Journal 100: 1, 178–181.

Laird, D.A. 2009. Impact of Biochar Amendments on Soil Quality for a Typical Midwestern Agricultural Soil. Presentation made at the North American Biochar Conference 2009, 9–12 August, University of Colorado at Boulder, USA. http://cees.colorado.edu/biochar_soils.html.

Lal, R. 2009. Challenges and opportunities in soil organic matter research in Eur. J. Soil. Sci. 60: 158–169.

Lane, M. 1980. The Story of the Steam plough Works: Fowlers of Leeds. Northgate Publishing Ltd, UK.

Leach, H.M.K. 1984. 1,000 Years of Gardening in New Zealand. Reed Publications, Wellington, New Zealand.

Lehmann, J., J. Gaunt, and M. Rondon. 2006. Biochar Sequestration in Terrestrial Ecosystems. Mitigation and Adaptation Strategies for Global Change 11: 403–427.

Lehmann, J. 2009a. 'Terra Preta Nova'—Where to From Here? pp. 473–486 in Amazonian Dark Earths: Wim Sombroek's Vision. Published by Springer Science, Netherlands.

Lehmann, J. 2009b. Testimony before the U.S. House Select Committee on Energy Independence and Global Warming. http://globalwarming.house.gov/files/HRG/061809agriculture/LehmannTestimony.pdf.

Lehmann, J., C. Czimczik, D. Laird, and S. Sohi. Stability of Biochar in the Soil. pp. 183–203. *In:* J. Lehmann, and S. Joseph [eds.]. 2009. Biochar Environmental Management, Science and Technology. Published by Earthscan, London, UK.

Lehmann, J., M.C. Rillig, J. Thies, C.A. Masiello, W.C. Hockaday, and D. Crowley. 2011. Biochar effects on soil biota—A review. Soil Biol & Biochem. 43: 1812–1836.

Levine, J.G. 2010. U.S.-Focused Biochar Report: Assessment of Biochar's Benefits for the United States of America, Published by The Center for Energy and Environmental Security and The United States Biochar Initiative, http://www.biochar-us.org/pdf%20files/biochar_report_lowres.pdf.

Macgregor, A.N. 2007. Can success be the cause of failure? J Organic Systems 2: 1.

MacKenzie, M.D. and T.H. DeLuca. 2006. Charcoal and shrubs modify soil processes in ponderosa pine forests in western Montana, Plant Soil 287: 1–2, 257–266.

Major, J. 2010. Guidelines on Practical Aspects of Biochar Application to Field Soil in Various Soil Management Systems, International Biochar Initiative, http://www.biochar-international.org/sites/default/files/IBI%20Biochar%20Application%20Guidelines_web.pdf.

Marris, E. 2006. Quoted J. Kimble re using no-tillage to sequester biochar, in pp. 624–626, Black is the New Green. Nature 442: 624–626.

McGonigle, T.P., M.H. Miller, and D. Young. 1999. Mycorrhizae, crop growth, and crop phosphorous nutrition in maize-soybean rotations given various tillage treatments, Plant Soil 210: 33–42.

Merfield, C.N. 2008. Trials of a crimper-roller for killing cover crops for organic and non-herbicide, no-till cropping. http://www.merfield.com/research/2009/trials-of-a-crimper-roller-for-killing-cover-crops-for-organic-and-non-herbicide-no-till-cropping.pdf.

Merfield, C.N. 2012. Biochar: The need for precaution? pp. 46–49 in March–April edition of Organic NZ, 71: 2. Published by Soil and Health Association of New Zealand Inc. Author-edited version of original article by the same title in Mother Earth, Spring 2011, the journal of Soil Association UK. http://www.merfield.com/research/2011/Biochar--the-need-forprecaution-2011-Merfield.pdf.

Miller, M.H. 2000. Arbuscular mycorrhizae and phosphorous nutrition of maize: a review of Guelph studies. Can J. Plant Sci. 80: 47–52.

Miller, M.H., T.P. McGonigle, H.D. and Addy. 1995. Functional ecology of vesicular-arbuscular mycorrhizas as influenced by phosphate fertilisation and tillage in an agricultural ecosystem. Crit Rev. Biotech. 15: 241–255.

Miller, R.M. and J.D. Jastrow. 1990. Hierarchy of root and mycorrhizal fungal interactions with soil aggregation. Soil Bio. Biochem. 22: 579–584.

Miller, R.M. and J.D. Jastrow. The Role of Mycorrhizal Fungi in Soil Conservation, pp. 29–44 *In*: G.J. Bethlenfalvay and R.G. Lindermann [eds.]. 1992. Mycorrhizae in Sustainable Agriculture, ASA Special Publication 54.

Miller, R.M. and J.D. Jastrow. Mycorrhizal fungi influence soil structure, pp. 3–18 in Y. Kapulnik. [ed.]. 2000. Arbuscular mycorrhizas: physiology and function. Kluwer, Dordrecht, The Netherlands.

Mitchell, J. and H. Mitchell. 2007. Te Tau Ihu o Te Waka—A History of Māori of Nelson and Marlborough, Volume 1—Te Tangata me Te Whenua—The People and the Land, Huia Publishers, Wellington, Aotearoa, New Zealand.

Moorfield, J.C. 2005. Te Aka—Māori-English, English-Māori Dictionary and Index. Published by Pearson Education, New Zealand. http://www.Māoridictionary.co.nz/.

Nishio, M. 1996. Microbial Fertilizers in Japan, Monograph, Extension Bulletin published by Food and Fertiliser Technology Center (FFTC), Taipei City, Taiwan.

Oades, J.M. 1993. The role of biology in the formation, stabilization and degradation of soil structure. Geoderma 56: 377–400.

Ocampo, J.A. and D.S. Hayman. 1981. Influence of plant interactions on vesicular-arbuscular mycorrhizal infections. II. Crop rotations and residual effects of non-host plants. New Phyto Vol. 87: 333–343.

Odum, E.P. 1983. Basic Ecology, 2nd Edn. Publ. By Holt-Saunders International Edition, Tokyo, Japan.

Odum, E.P. and G.W. Barrett. 2005. Fundamentals of Ecology (5th Edition), Publ. by Thomson, Brooks Cole, USA.

Ogawa, M. and Y. Okimori. 2009. Pioneering Works in Biochar Research, Japan. Aust. J. Soil Res 48: 489–500.

Pietikainen, J., O. Kikkila, and H. Fritze. 2000. Biochar as a habitat for microbes and its effect on the microbial community of the underlying humus. Oikos 89: 231–242.

Poincelot, R.P. and D. Bennett. 1988. Organic No Dig, No Weed Gardening: A Revolutionary Method for Easy Gardening. Published by Thorsons, UK.

Rees, W.E. 2010. Thinking Resilience. pp. 25–40, in The Post Carbon Reader—Managing the 21st Century's Sustainability Crises, Edited by Heinberg R. and Lerch D., Publ. by Watershed Media, Healdsburg, California, USA.

Ribeiro, F., S.E. Justice, P.R. Hodd, and C.J. Baker. No-tillage Drill and Planter Design—Small Scale Machines. pp. 204–225. *In:* C.J. Baker, K.E. Saxton, W.R. Ritchie, W.C.T. Chamen, D.C. Reicosky, M.F.S. Ribero, S.E. Justice, and P.R. Hobbs. 2006. No-Tillage Seeding in Conservation Agriculture, 2nd Edition, Published by F.A.O. and CAB International.

Rillig, M.C. and D.L. Mummey. 2006. Mycorrhizas and Soil Structure. Tansley Review in New Phytol 171: 41–53.

Ritchie, B., C.J. Baker, and M. Hamilton-Manns. 2000. Successful No-tillage in Crop and Pasture Establishment, Centre for International No-tillage Research and Engineering, Fielding, New Zealand.

Saito, M. 1990. Charcoal as a micro-habitat for VA mycorrhizal fungi, and its practical application. Agriculture, Ecosystems and Environment 29: 341–344.

Saito, M. and T. Marumoto. 2002. Inoculation with arbuscular mycorrhizal fungi: the status quo in Japan and future prospects. Plant Soil 244: 273–279.

Seybold, C.A., J.E. Herrick, and J.J. Brejda. 1999. Soil Resilience: A fundamental component of soil quality, Soil Science. 164: 224–234.

Simpson, R.T., D.F. Serita, J. Six, and R.K. Thiet. 2004. Preferential accumulation of microbial carbon in aggregate structures in no-tillage soils, in Soil Sci. Am. J. 68: 1249–1255.

Six, J., E.T. Elliott, and K. Paustian. 2000. Soil macro aggregate turnover and micro aggregate formation: a mechanism for C sequestration under no-tillage agriculture. Soil Biol. Biochem. 32: 2099–2103.

Six, J., C. Feller, K. Denef, S.M. Ogle, J.C. de Moraes, and A. Albrecht. 2002. Soil organic matter, biota and aggregation in temperate and tropical soils—Effects of no-tillage (review article) Agronomie 22: 755–775.

Six, J., H. Bossuyt, S. Degryze, and K. Denef. 2004. A history of research on the link between (micro) aggregates, soil biota, and soil organic matter dynamics, Soil Tillage Res. 79: 7–31.

Smith, S.E. and D.J. Read. 2008. Mycorrhizal Symbiosis (3rd Edition), Publ. by Academic Press, Elsevier, New York, USA.

Sohi, S.P., E. Krull, E. Lopez-Capel, and R. Bol. 2010. A Review of Biochar and Its Use and Function in Soil, Advances in Agronomy 105: 47–82.

Solaiman, Z.M., P. Blackwell, L.K. Abbott, and P. Storer. 2010. Direct and residual effect of biochar application on mycorrhizal root colonisation, growth and nutrition of wheat, Aust. J. Soil. Res. 48: 546–554.

Sparling, G., R.L. Parfitt, A.E. Hewitt, and L.A. Schnipper. 2003. Ecological risk assessment —Three approaches to define soil organic matter contents, in J. Environ Qual. 32: 760–766.

Steiner, C. Biochar in agricultural and forestry applications, in Biochar From Agricultural and Forestry Residues—A Complimentary Use of "Waste Biomass" pp. 1–14. *In:* Levine [ed.]. 2010. U.S.-Focused Biochar Report: Assessment of Biochar's Benefits for the United States of America, Published by The Center for Energy and Environmental Security and The United States Biochar Initiative, http://www.biochar-us.org/pdf%20files/biochar_report_lowres.pdf.

Steiner, C., W.G. Teixeira, W.I. Wood, and W. Zech. 2009. Indigenous knowledge about 'Terra Preta' formation. pp. 193–204 in Amazonian Dark Earths: Wim Sombroek's Vision, Published by Springer Science, Netherlands.

Stout, R. 2011. Gardening Without Work. Published by Norton Creek Press, USA.

Stuart, A. 2012. Biochar: An ancient tool for a more sustainable future? in March–April edition of Organic NZ, 71: 2. 44–45, Published by Soil and Health Association of New Zealand.

Thies, J.E. and M.C. Rillig. 2009. Characteristics of Biochar: Biological Properties. pp. 85–105 in Biochar For Environmental Management, Science and Technology, Published by Earthscan, London.

Tisdall, J.M. and J.M. Oades. 1982. Organic matter and water-stable aggregates in soils. J. Soil Sci. 33: 141–163.

Urbanaska, K.M., N.R. Webb, and P.J. Edwards. 1997. Restoration Ecology and Sustainable Development. Publ. by Cambridge University Press, UK.

van der Heijden, M.G.A. and I.A. Sanders. 2002. Mycorrhizal Ecology, Ecological Studies 157, Published by Springer-Verlag, Berlin.

van der Heijden, M.G.A., R.D. Bardgett, and N.M. van Straalen. 2008. The unseen majority: soil microbes as drivers of plant diversity and productivity in terrestrial ecosystems. Ecology Letters, 11: 296–310.

Verheijen, F., S. Jeffrey, A.C. Bastos, M. van der Velde, and I. Diafas. 2010. Biochar Application to Soils—A critical scientific review of effects on soil properties, processes and functions. Publ. by European Commission, Joint Research Centre, Institute for Environment and Sustainability.

Waisel, Y., A. Eshel, and L. Kafkafi. 2002. Plant Roots—The Hidden Half (3rd Edition), Publ. by Marcel Dekker Inc. New York, USA.

Walker, B. and D. Salt. 2006. Resilience Thinking: Sustaining Ecosystems and People in a Changing World. Published by Island Press, Washington DC, trademark of The Centre for Resource Planning.

Wang, H., L. He, and K. Lu. 2012. Biochar for remediation of soils contaminated by organic pollutants, presented at Biochar Workshop, Feb. 9–10, Massey University, Palmerston North, New Zealand.

Wardle, D.A., O. Zackrisson, and M.C. Nilsson. 1998. The charcoal effect in Boreal forests: mechanisms and ecological consequences. Oecologia 115: 419–426.

Wardle, D.A. 2002. Communities and Ecosystems—Linking the Aboveground and Belowground Components. Monographs in Population Biology 34, Princeton University Press, New Jersey, USA.

Warnock, D.D., J. Lehmann, T.W. Kuyper, and M.C. Rillig. 2007. Mycorrhizal responses to biochar in soil—concepts and mechanisms, Plant Soil 300: 9–20.

Williams, M.M. and J.C. Arnott. 2010. A comparison of variable economic costs associated with two proposed biochar application methods, Ann. Environ. Sci. 4: 23–30.

Woolf, D. 2008. Biochar as a soil amendment: A review of the environmental implications http://www.orgprints.org/13268/1/Biochar_as_a_soil_amendment—a_review.pdf.

Wright, S.F., J.L. Starr, and I.C. Paltineanu 1999. Changes in Aggregate Stability and Concentration of Glomalin during Tillage Management Transition, Soil Sci. Am. J. 63: 6 1825–1829.

Index

About the Editors

Dr. Natalia Ladygina graduated from the Department of Biology of Kuban State University, Krasnodar, Russia as a microbiologist. Thereafter she worked as a researcher at Institute of Biochemistry and Physiology of Microorganisms of Russian Academy of Sciences in Pushchino, Moscow region, Russia. As a Marie Curie fellow she obtained her Ph.D. in Biology from Lund University, Sweden. Her work has focused on trophic interactions among different soil organisms in grassland soils.

Dr. Francois Rineau has a Ph.D. in Microbial Ecology obtained at Nancy University, France. He worked at the Microbial Ecology Lab in Lund University, Sweden; currently he is at Environmental Biology department of Hasselt University, Belgium. His research projects are focused on ectomycorrhizae: from their community ecology to their metabolism, their physiology and their strategies of nutrient assimilation.

Color Plate Section

Chapter 6

Amorphous Carbon Matrix with nano sized mineral phases and micro and mesopores

Graphitic Carbon with interstitial metal atoms

$CaCO_3 +/- MgO$

CaH_2PO_4

Dislocations

Macropores

Clay particles

Fe Oxide or Sulphide

KCl

Silica

TiO_2

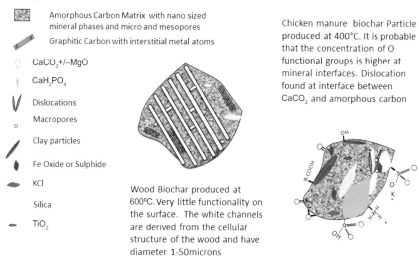

Chicken manure biochar Particle produced at 400°C. It is probable that the concentration of O functional groups is higher at mineral interfaces. Dislocation found at interface between $CaCO_3$ and amorphous carbon

Wood Biochar produced at 600°C. Very little functionality on the surface. The white channels are derived from the cellular structure of the wood and have diameter 1-50microns

Figure 1. Schematic of complex structure of low temperature biochars.

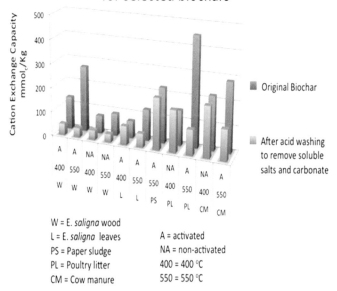

Mean Cation Exchange Capacities for selected biochars

Legend:
- Original Biochar
- After acid washing to remove soluble salts and carbonate

W = E. saligna wood
L = E. saligna leaves
PS = Paper sludge
PL = Poultry litter
CM = Cow manure

A = activated
NA = non-activated
400 = 400 °C
550 = 550 °C

Figure 2. Plot of CEC versus feedstock, temperature and processing technique (Plot of data presented by Singh and Singh (2010).

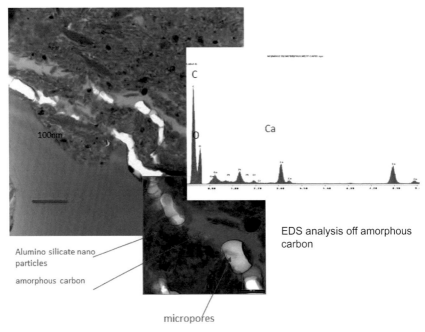

Alumino silicate nano particles

amorphous carbon

micropores

EDS analysis off amorphous carbon

Figure 3. Transmission Electron Microscopy (TEM) image and Energy Dispersive Spectroscopy (EDS) spectrum of interface between amorphous carbon and organomineral phase in a Terra Preta particle.

Figure 4. a. Aging bamboo and wood in a pond; b. Burning aged wood and bamboo in open fire; c. Biochar from aged wood produced in open fire on field.

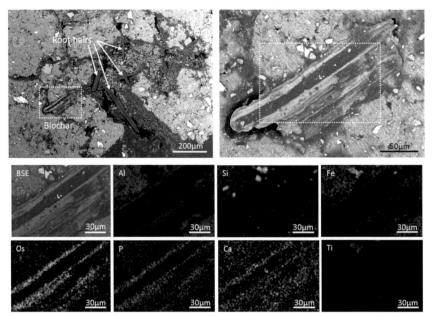

Figure 5. Aged biochar with root hair and microorganisms inside pore. Source: Electron Microscope Unit, University of NSW.

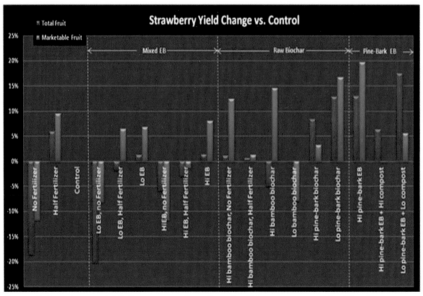

Figure 9. Strawberry yield changes as a function of soil treatments and fertilization rate. All blocks received the full complement of fertilizer except where labeled "No Fertilizer" or "Half Fertilizer." "Hi" and "Lo" refer to the treatment concentrations as detailed in Table 1. The error bars in these results are approximately ±5% for the Controls and mixed BMC data and ±10% for the raw biochars.

Chapter 7

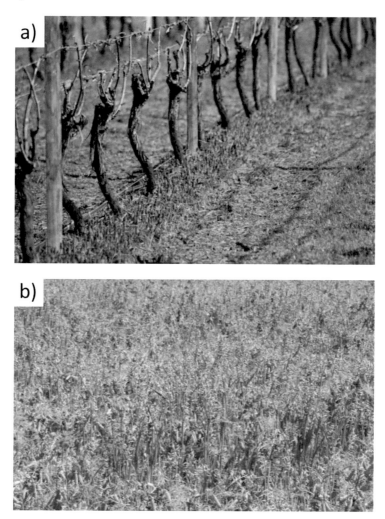

Figure 1. Examples of soil conservation methods: a) Grape Hyacinth groundcover growing in understory of vineyard. b) Peas and oats green manure crop.

Figure 16. Close up view of biochar feedstock *Pinus radiata* sawdust compressed pellets (left), charred pellets (centre), and clay-coated biochar pellets (right).

For Product Safety Concerns and Information please contact our EU
representative GPSR@taylorandfrancis.com Taylor & Francis Verlag GmbH,
Kaufingerstraße 24, 80331 München, Germany

Printed and bound by CPI Group (UK) Ltd, Croydon, CR0 4YY

01/05/2025

01858617-0002